The Rediscovery of the Wild

The Rediscovery of the Wild

edited by Peter H. Kahn, Jr., and Patricia H. Hasbach

The MIT Press
Cambridge, Massachusetts
London, England

MIT Press books may be purchased at special quantity discounts for business or sales promotional use. For information, please email special_sales@mitpress.mit.edu or write to Special Sales Department, The MIT Press, 55 Hayward Street, Cambridge, MA 02142.

This book was set in Sabon by the MIT Press. Printed on recycled paper and bound in the United States of America.

Library of Congress Cataloging-in-Publication Data

The rediscovery of the wild / edited by Peter H. Kahn, Jr., and Patricia H. Hasbach.
    p.   cm.
Includes bibliographical references and index.
ISBN 978-0-262-01873-9 (hardcover : alk. paper) — ISBN 978-0-262-51833-8 (pbk. : alk. paper)
1. Human beings—Effect of environment on. 2. Human ecology—Philosophy. 3. Philosophy of nature. 4. Mind and body. 5. Wilderness areas. I. Kahn, Peter H. II. Hasbach, Patricia H.
GF51.R44   2013
304.2—dc23
2012025616

10   9   8   7   6   5   4   3   2   1

To the mystery of the wild

Contents

About the Contributors

E. N. Anderson is a professor emeritus of anthropology at the University of California at Riverside, and a researcher who studied ethnobiology, cultural ecology, political ecology, and medical anthropology in Hong King, British Columbia, California, and the Yucatan Peninsula in Mexico. His most recent book is *The Pursuit of Ecotopia* (2010).

G. A. Bradshaw is founder and executive director of the Kerulos Center. Her most recent book is *Elephants on the Edge: What Animals Teach Us about Humanity* (2009).

Cristina Eisenberg is an Oregon State University scientist who studies how wolves affect ecosystems throughout the American West. Her book *The Wolf's Tooth: Keystone Predators, Trophic Cascades, and Biodiversity* was published in 2010.

Dave Foreman is a longtime environmental activist with a focus on landscape-scale conservation in North America. He is the director of the Rewilding Institute and author of several books, including most recently *Man Swarm and the Killing of Wildlife* (2011).

Patricia H. Hasbach is a psychotherapist in a private practice in Eugene, Oregon, and a faculty member at Lewis and Clark College in Portland, Oregon, and Antioch University Seattle. She is a coeditor of *Ecopsychology: Science, Totems, and the Technological Species* (2012).

Peter H. Kahn, Jr., is a professor in the Department of Psychology as well as the director of the Human Interaction with Nature and Technological Systems (HINTS) Lab at the University of Washington. He is the author of *Technological Nature: Adaptation and the Future of Human Life* (2011).

Ian McCallum is a medical doctor, analytic psychologist, psychiatrist, and adjunct faculty member at the Nelson Mandela Metropolitan University Business School in Port Elizabeth, South Africa. He is the author of two anthologies of poems and *Ecological Intelligence: Rediscovering Ourselves in Nature* (2008).

Gail F. Melson is a professor emeritus in the Department of Human Development and Family Studies at Purdue University. She is the author of *Why the Wild Things Are: Animals in the Lives of Children* (2001).

Bridget Stutchbury is a professor in the Department of Biology at York University in Toronto, where she holds a Canada Research Chair in Ecology and Conservation Biology. Her most recent book is *The Private Lives of Birds* (2010).

Elizabeth Marshall Thomas is a writer. She is the author of *The Old Way: A Story of the First People* (2006), about Ju/wa Bushmen during the 1950s, when they lived as hunter-gatherers.

Jack Turner taught philosophy at the University of Illinois and is a visiting scholar in the environmental humanities graduate program at the University of Utah. He was a longtime mountaineering guide in the Tetons, South America, and the Himalayas, and his most recent book is *Travels in the Greater Yellowstone* (2008).

Introduction

Peter H. Kahn, Jr., and Patricia H. Hasbach

Many people who currently advocate for the importance of nature in human lives focus on what's close at hand: domestic, nearby, everyday nature. It might be a local park, one's own garden, one's dog, a nearby walking trail, or birds finding sustenance in urban feeders. But domestic nature is only part of what we need. The other part is wild nature. For as a species we came of age in a natural world far wilder than today, and much of the need for wildness still exists within us, body and mind.

Wildness often involves that which is big, untamed, unmanaged, not encompassed, and self-organizing, and unencumbered and unmediated by technological artifice. We can love the wild. We can fear it. We are strengthened and nurtured by it. Part of what the wild in the natural world offers are clear choices, with the sanity that emerges through outcomes that can involve warmth and cold, a full belly, the strength to move, along with the freedom and land to do so, and living and dying. Is it safe to cross the river or not? Should I set up camp now or can I extend the day's climb a little farther? I see the grizzly and the grizzly sees me, which, as Jack Turner (1996, p. 85) says, is a wild experience:

In a most intimate, carnal way, an experience that is marked by gross alterations in attention, perception, body language, body chemistry, and emotion. Which is to say you feel yourself as part of the biological order known as the food chain, perhaps even as part of a meal.

More than sixty years ago, Aldo Leopold (1949/1970, p. 133) wrote:

We all strive for safety, prosperity, comfort, long life, and dullness. The deer strives with his supple legs, the cowman with trap and poison, the statesman with pen . . . but too much safety seems to yield only danger in the long run. Perhaps this is behind Thoreau's dictum: In wildness is the salvation of the world. Perhaps this is the hidden meaning in the howl of the wolf, long known among mountains, but seldom perceived among men.

Most of us have gotten soft. This is true for many of us who live in urban settings, and environmentalists and nature lovers as well. There are fewer wild lands. There is less wild nature. And thus people have fewer and fewer opportunities to interact with the wild. In turn, as new generations come of age in diminished wild landscapes, their baseline shifts for what counts as wild nature. Kahn (1999, 2011) speaks of this problem in terms of environmental generational amnesia: that members of each generation construct their conception of what is environmentally normal based on the natural world they encounter in their childhood. The crux is that with each ensuing generation, the amount of environmental degradation and loss of wildness increase, but each generation conceives of that new condition as nondegraded and normal. It is not simply that we adapt to diminished nature, don't know it, and do fine. Rather, we adapt, don't know it, and unknowingly do not flourish as individuals and a species. By analogy, imagine that we were born in San Quentin and had no conception of what lay outside the prison walls. Most of us would live full lives as measured by biological years. But they would not be full as measured by standards based on who we are as a species, shaped by hundreds of thousands of years of evolution in a natural world. As we lose the wild, we diminish ourselves.

One of the challenges and opportunities for our age is to describe, name, and bring forward wild nature, so that we can more deeply engage what's left and, where possible, recover what's lost. That is the charter of this edited volume, which the authors contribute to with grounded experience, psychological depth, philosophical sophistication, and lovely writing.

In chapter 1, "Quantifying Wildness: A Scientist's Lessons about Wolves and Wild Nature," Cristina Eisenberg examines the paradox of her work: that even as she studies wildness through the lens of empirical science, her experience of the wild's mystery deepens. As a conservation biologist and researcher, Eisenberg places radio-collars on one of the wild's most evocative animals—the wolf—and justifies doing so "because that is what it takes to obtain data that will help sustain wolves as part of society's attempt to manage wildness." After many years in the field, she recognizes the importance of the relationship she forms with each of the animals she studies: "The wolves oblige me by allowing themselves to be trapped and collared. We have a tacit understanding: I handle them with the highest care and respect; they provide a wealth of data that could not be obtained otherwise." With unconventional candor and openness, Eisenberg shares stories of the Belly River wolf pack she studies and also of her encounter with a dangerous grizzly bear. She reflects on the lessons wild nature has for all of us. If we can heed these lessons, she thinks it's possible that "we can relearn how to deal with fear and wildness . . . [and] even learn how to mend the damage we've done to the earth."

In chapter 2, "The Wild and the Self," Turner invites the reader to think anew about wildness. He says it's not just out there in wild animals, big lands, and uncharted territories, which was the focus of much of his earlier writing. Rather, Turner now draws on the Chinese term *tzu-jan*—a concept of the self—which he defines as "the spontaneous newness that pours forth each moment, in the world and our experience." In this way, wildness exists in varied forms, "from the chemical elements we inherited from exploding stars, and that still course through our veins, to Gaia, the functioning of our immune system, and the stalking bacteriophage . . . from the trillions of stars down to the trillions of synapses." Drawing on pop culture, poetry, literature, etymology, physics, and Chinese philosophy, Turner fleshes out the unfolding, dynamic, and constant presence of wild nature, and highlights our embeddedness in it and being "of it." He calls this pervasive wildness the Vast Array, and asks why we aren't collectively stunned with appreciation. Turner believes that until we elaborate a rich appreciation of wildness, and have wonder, awe, and love for the wild, political solutions—including protecting biodiversity, wildlife corridors, and large preserves of land—won't work. He concludes that however abstract the ultimate analysis of wild nature, we can only experience its spontaneous presence, and to do so, we must, like those before us, become intimate with natural history and spend time in wild places.

One of the central assumptions of many people who advocate for wildness in our lives today is that patterns of interaction with wild nature lie deep within our genetic history. But how would we know how people actually interacted with nature tens of thousands of years ago? One answer is to draw on accounts from the 1950s of indigenous African people, the Bushmen, before they had had virtually any contact with the West. These are a people who had lived sustainably in this African landscape for well over thirty-five thousand years, with a lifestyle that presumably changed little during that time. Thus to understand them then, in the 1950s, is to gain access, in a way that bones and fossils cannot render, to aspects of who we were at an earlier time in our evolutionary history.

Elizabeth Marshall Thomas provides us with such access in chapter 3, "The Old Rules." When she was nineteen years old, she and her father, mother, and brother ventured into the Kalahari Desert, and eventually made contact with the Ju/wasi, one of the five groups of Bushmen that inhabited that area. Between 1950 and 1955, the Marshalls made three expeditions into this area. In 2006, Thomas wrote of her experiences with the Ju/wasi in an important book, *The Old Way*. In her chapter here, she expands on some of the themes from this book. She writes, for example, of her encounter with a leopard one night, and how the animal followed an old rule: avoid unnecessary confrontations. The Ju/wasi hunted with a passion, killed wild animals, and on occasion fended off predators, but they knew this old rule, too, even among themselves. Thomas observes:

Ju/wa culture contained multiple, redundant ways of cementing personal relationships and keeping the many small groups united. . . . Ill will and jealousy were considered to be dangerous, and the people did everything they could to avoid these threats to their unity. The "fittest" people were those who felt the importance of this most strongly.

Thomas unites many details of Ju/wa culture and knowledge of the wild with what we need, still today, to flourish as a species.

In "Wild Wings," chapter 4, field biologist and ornithologist Bridget Stutchbury describes how the world of birds can be a portal to wildness for the forty million bird watchers in the United States—but only if we know how to truly see. Stutchbury writes, "Sex, adultery, divorce, and daily threats and acts of violence are common in the bird world that lives in the backyard." She argues that a deep knowledge and understanding of the natural history and behavioral ecology of birds provides us with a more meaningful experience than simply seeing birds as pretty ornaments at our backyard feeders. She calls for a nature literacy that will not only enrich our experience of nature but also be a powerful tool for conservation and environmental sustainability. Stutchbury discusses how human-driven changes in the global landscape are domesticating wild birds, selecting for generalists over specialists, and shifting the baseline of what we consider wild. She cautions that as we lose wild places through actions such as clear-cutting a forest, we lose not only the plants and animals that make up that habitat but evolutionary history as well. Stutchbury concludes that many bird species are evolving in response to human influences; thus there are few species that are fully wild, and what we actually observe could be "an artifact of human changes to the natural world."

There is clear evidence that speaks to the educational and therapeutic significance of pets in children's lives, but what about the role of wild animals? Are they relevant, too, especially given that most children grow up in urban settings? Gail F. Melson explores this question in chapter 5, "Children and Wild Animals," drawing from a wide range of disciplines, including psychology, sociology, animal science, environmental science, geography, and human-animal interaction. She finds that children today are drawn to wild animals in such settings as zoos, aquariums, butterfly farms, and bird sanctuaries, and interact with wild animals they find close at hand, such as wild birds that visit backyard bird feeders, squirrels and rabbits in a park, or frogs, ants, and spiders. Melson argues that children's engagement with wild animals is rooted in human evolutionary history, and then, through historical analyses, she shows how culture can shape a child's understanding of and engagement with the wild. Perhaps most important, Melson brings together evidence to suggest that engagement with wild animals helps to foster children's perceptual, cognitive, self, social, emotional, and moral development. Wild animals, according to Melson, "reflect and refract the self, act as social others, and prompt moral reasoning about other species and one's place in the universe." Children need wild animals in their lives.

Adults do, too. But oddly our behavior toward wild animals is often far from benign. In chapter 6, "Living Out of Our Minds," G. A. Bradshaw helps us to see some of the horrible abuses we perpetrate on wild animals. There's Bradshaw's account of the African elephant Medundamelli, captured at two, and years later, during her "training" at the San Diego Wild Animal Park and Zoo, "the head zoo elephant keeper and four other men sought to 'discipline' the female elephant. While all four of her legs were chained to the extent that she was nearly prone, the men beat her with wooden ax handles." There's Bradshaw's account of Jeannie, a chimpanzee, captured from the African wilderness, who endured decades of brutalization as a biomedical subject, including multiple cervical, liver punch, wedge, and lymph node biopsies; infection with HIV as well as hepatitis NANB and C virus; and being subjected to over two hundred "knockdowns" (anesthetization by a dart gun). On a beautiful note, Bradshaw demonstrates that both abused animals and abused humans (such as those who have survived concentration camps) have within them the capacity to choose life—to move outside oneself with compassion and gifts to others. Bradshaw contends that wild animals undergo physical and psychological trauma much as humans do. She says that to think otherwise is to be trapped by an outdated mechanistic conception of consciousness and life. Here Bradshaw draws on cutting-edge research in the neurosciences to show support for a "new field of trans-species psychology," in which "the study of human and other animal minds and souls" merges.

In "A Wild Psychology," chapter 7, Ian McCallum says he knows what he wants to do with the rest of his life: be a voice for the wild places and the wild part of the human psyche. He believes we are in danger of losing not only wild places but also the wild landscapes of the soul that form our human identity. McCallum identifies one burning question regarding the human contribution to our environmental predicament, "*Why, as a species, do we behave in such a destructive way in the first place?*" He answers that we don't want to face up to our animal, evolutionary, and wild origins. McCallum argues that we fear wildness too much, erroneously seeing anything that is spontaneous or indifferent to human existence as threatening. What we don't understand, we fear, and we ultimately destroy what we fear. McCallum says we must rediscover and understand the wild tokens in ourselves "at the biological, social, and psychological levels of human existence, and that we protect them . . . fiercely." He also points out three survival-oriented intelligences in humans: a reactive, primal intelligence; a responsive, social intelligence; and a reflective, self-aware intelligence. It is the combination of these intelligences that allow us the free will—the deliberate and conscious capacity—to say "no." Evolutionary thinking allows him to see human behaviors in a different light, and helps him to recognize that psychological healing will never be complete as long as it is seen as separate from the world and web of life. To become aware of this connection and then to begin to live it will add depth to our sense of belonging, and will "go a long way toward a more practical and genuine reappraisal of our place and position on earth."

In chapter 8, "Culture and the Wild," E. N. Anderson deals thoughtfully with what must be apparent to many readers: that conceptions and relationships with wildness are substantively shaped by culture. According to Anderson, for example, "Native Americans and early Celtic peoples saw the relatively wilder areas of the world as both good and important: realms of hunting, gathering, and spiritual activities." In contrast, historically the Chinese have seen wild areas in two ways: as an area to farm and from which to extract natural resources, and as an area for self-cultivation and renewal. Anderson provides his own unique and stunningly beautiful translations of some Chinese nature poems. For instance, here's one from around the eighth century: "Thousand mountains no bird flies / Ten thousand paths no human track. / Solitary boat old man, split-bamboo raincloak, / Fishing, alone cold river snow." The construct of culture, notes Anderson, refers to a repository of ideas and representations, which then serve as sources of guidance. In this way, culture is not in itself homogeneous. Using the Bible as an example, Anderson points out that it embodies two contradictory cultural convictions. One, represented in the first chapter of Genesis, gives humans full dominion over nature. But "the second chapter contains an entirely different creation story that makes people the stewards of God's landscape, not the dominators."

In contrast to those like Dave Foreman in the following chapter, Anderson is highly suspect of wild areas that are set apart from people, especially indigenous people, who live on these lands and tend to manage them well. That said, Anderson understands that the "idea of people-in-nature is hard to establish when people are in fact living in cities." Thus, Anderson argues for educational programs and environmental management policies that are grounded in direct experience with nature. The reality of nature itself can be seen to check a relativism in Anderson's account that is often present in other environmental cultural theorists.

In "Five Feathers for the Cannot Club," chapter 9, Dave Foreman begins by asking, "How do we fit in and live with all the other earthlings for the long haul?" He believes that we must acknowledge all other earthlings as kindred and learn to be good neighbors in this community of earth. According to Foreman, being a good neighbor means recognizing the intrinsic value of all beings, and being willing to brake "our self-willed might." Foreman contends that shielding evolution is the true work of conservation. How is this best done? His answer is through keeping wilderness and wildlife free. Foreman provides a historical account of the conservation movement and discusses two kinds of conservationists. One he calls resourcists, who "believe in managing natural resources for products and uses for humans." The other he calls conservationists, who "believe in keeping things wild for their own sake." Critical of the former, Foreman draws on characteristics of five common bird species to highlight what conservationists—members of the Cannot Club—can do to keep wild places and wild beings free, and for the conservation movement to succeed.

Finally, in chapter 10, "The Rewilding of the Human Species," we suggest that our Paleolithic ancestors lived a life more wild than all of us do today, and that much of that wildness still exists within the architecture of our bodies and minds. Accordingly, we contend that for people to flourish now, much of that wildness needs to be rediscovered, re-engaged, developed, and lived. Hence we argue for human rewilding. In so doing, we develop five central ideas. First, we reinstate the importance of the primal self not only in relation to a wild nature "out there" but also within, including sexuality and aggression. Second, we suggest that humans fear wild nature too much. Third, we maintain that humans need to welcome greater variation and periodicity in the satisfaction of our needs and desires. Fourth, we show how rewilding will occur more fully within flatter social hierarchies, which minimize control and domination. We develop this idea by examining patriarchal systems that have existed in many cultures since the rise of agriculture and have been reified through the emergence of what we term the warrior ethos. In other words, through a similar genesis, we closely link the domination of the wild with the domination of one people over another. Fifth, we recognize that as a species, while we are still connected to our Paleolithic selves, we have also adapted to modern times, becoming increasingly technological and scientific. We therefore argue that by rewilding ourselves, and then integrating many of the "original" and "primal" ways of knowing and being into our technological selves and scientific culture, we can, in principle, have it better than we have ever had it before. In this sense we offer a vision at once practical, immediately applicable, and utopian in scope.

These ten chapters help to establish the central proposition of this volume: as individuals and a species, we need to rediscover the wild. The chapters show how to do so.

Building on the content of these chapters, there appear to us five overarching issues that stand out in some relief against this intellectual landscape. These issues are worth characterizing, especially insofar as they have the potential to help shape the direction of scholarship over the decade ahead.

1. Is the Wild Everywhere?

In the edited volume *Children and Nature*, naturalist Robert Michael Pyle (2002, p. 306) wrote a chapter titled "Eden in a Vacant Lot," in which he argued that "nothing is less empty to a curious, exploring child than a vacant lot, nothing less wasted than waste ground, nothing more richly simmered in promise than raw ground." Pyle would likely say that wild nature exists in the vacant lot, too. After all, look at a weed growing up through the asphalt. The weed embodies an autonomous, self-willed, natural entity that is exerting its natural course for life independent of human will—Eden in a vacant lot; the wild weed. Turner (this volume) argues for an even stronger position when he expresses frustration with people who seek wildness predominantly in canonical wild areas:

The idea that we have to go to Yellowstone to find or experience the wild is absurd. We simply must learn to switch scales with ease, to move deftly from the vastly big to the infinitesimally small, integrated hierarchies. . . . Wildness is not limited to the human scale, our love for large fauna, and what we can see. We need to get over that. Study everything from quarks to worms to the clusters of galaxies portrayed in the Hubble Deep Field images; everything we can find, everywhere, is coming forth, spontaneously, moment by moment.

Thus Turner says "we need to get over" privileging the canonical wild. For Turner, the wild exists in "everything," and is especially present on scales larger and smaller than the human senses can directly perceive.

In our view, while human scale isn't everything, it's a lot. Human beings are embodied in a physical form and came of age enacting wild patterns on human scales. For example, we ran down animals, such as the bull eland, in the African heat. Our running might have lasted for four or six hours. There are optimal distances today for human long-distance running that match this wild pattern. We ran after animals that had a large enough size to justify such an expenditure of energy. We could also be prey to large animals, especially the lion, and we modified our behaviors accordingly. For instance, we walked the savanna during the heat of the day, when the lions rested. It's not an accident therefore that people are presently drawn to the large animals in zoos, or pay large sums of money to experience the wildlife on African safaris. These animals mattered to us in our evolutionary history.

It's also the case that we apprehend the world not just through our mind but also through our senses, which require a material world that can be sensed. Can we hear, taste, smell, and touch a star exploding in the distant past, whose faintest radiation only now reaches us? Can a prisoner access wild nature by appreciating that star's radiation, the weed that grows spontaneously in the prison yard, or the bacteria in the stomach of his cell mates? By all means, there is value in appreciating the wild in such ways. But if we give up the canonical wild, which exists on a human scale, we'll justify its destruction even more. Then there will be no more old-growth forests, no more open lands where one can hike for days or weeks in a single direction and not meet the other side, and no more animals that can check our hubris by their own large powers of volition—whales, porpoises, turtles bigger than we can scope in our arms, elephants that know a vast range walking, and bears that amble with impunity.

2. Does Technology Diminish the Experience of the Wild?

Many homes have gas fireplaces that technologically mimic a wood fireplace, which itself brings the primal fire circle into a built domicile. Many people have grown up with television programs about nature, such as with Animal Planet and Discovery Channel. Webcams provide real-time images of animals in zoos and nature areas throughout the world. People play with robotic dogs and even find companionship with them. People also interact with simulated wild animals. For example, in the promotional material for the interactive digital game Kinectimals, we are told that by playing this game you "get in touch with your wild side," and that it "invites children, their parents and animal lovers of all ages to build lasting friendships with some of the world's most exotic creatures." It's not just wild animals that exist virtually. People also farm in virtual environments such as Farmville. All of these technologies fit under what Kahn (2011) refers to as *technological nature*: technologies that mediate, augment, or simulate the natural world. Kahn and his colleagues investigated the psychological effects of interacting with technological nature. In chapter 10 (this volume), we report on some of these studies. This body of research shows that experiencing technological nature is better for people than not experiencing any nature, but not as good as actual nature.

That said, technologies provide access into wild areas. A pickup truck provides access to wilderness trailheads. A pair of hiking boots provides a means to walk those lands. Technologies can also increase access to wildness beyond the human senses. This was Turner's point above, when he said that technologies allow us to study "everything from quarks to worms to the clusters of galaxies portrayed in the Hubble Deep Field images." Closer to home, as Stutchbury (this volume) shows, binoculars offer millions of birders visual access to a form of wild nature that they would be unable to see unaided. Scuba gear similarly provides divers a means to access the depths of the wild ocean. And mountaineering gear allows climbers to summit the highest mountains of the world.

But there is a slippery slope here that needs to be called out now, for if left unchecked it will hugely affect people's interactions with the wild in the decades ahead. Imagine that we could make bifocal contact lenses that functioned in part as binoculars. We could then walk the birding lands and immediately switch to and from binocular vision. Let's next embed night-vision lenses into these contact lenses, and then fix those lenses permanently into the eye. Now we can effortlessly walk at night and see more than we have ever seen. Next, let's embed permanently into our ears a (future) technological device that allows us to hear a hundred times more sensitively and with a tenfold greater range of frequencies. We already have artificial hearts and limbs. Imagine the time when these technological devices are stronger than their biological counterparts. They will provide greater access into the wild. Maybe everyone would then be able to hike up to Half Dome in a day or reach Mount Everest's summit in a week. We are on our way to becoming cyborgs: biological beings fused with technological devices, and kept alive by technological machinery embedded within us and off-loaded through wireless connections to servers in distant locations. Does all this technology connect us better to the wild? We have our doubts. For the wild means in part the natural, which means in part unencumbered by technological human artifice and outside human control. The more we use technology to access the wild, the more we may gain, but also the more we will lose.

3. Self-regulation as a Defining Characteristic of Wildness

For many people, for many centuries, the wild has conjured images of something fearful "out there" that needs to subjugated. As the refrain in the *Wizard of Oz* goes, "Lions, and tigers, and bears, oh my"—dangerous animals all. Inner wildness has been viewed with a similar orientation. Some people believe, for example, that children are born essentially wild, and through the process of parenting and schooling need to become civilized (Rafferty, 1970). In our view, however, all such characterizations of outer and inner wildness fundamentally miss a defining characteristic of it: self-regulation.

What we have in mind here can be illustrated by unpacking what the psychologist Lawrence Kohlberg (1969, 1984) embedded in his moral developmental stage theory, especially in terms of what he meant by autonomy. Based on decades of empirical studies, Kohlberg proposed that people undergo a developmental progression in their understanding of morality, which encompasses six developmental stages. Each stage, according to Kohlberg, hierarchically integrates an earlier stage in its transformation. What is critical for our discussion is what Kohlberg meant by the idea of autonomy, which grounds stages five and six. On the one hand, Kohlberg does mean increasing independence from other people (stages one to two) and from conventional expectations (stages three to four). On the other hand, Kohlberg didn't mean autonomy as a divisive individualism but rather as highly social, developed through reciprocal interactions on a microgenetic level, and evidenced structurally in incorporating and coordinating considerations of self, others, and society (stages one through four). That incorporation and coordination is fundamental to the idea of a later stage being a hierarchical transformation of earlier ones.

In his book *The Abstract Wild*, Turner (1996, p. 113) says something similar:

Although autonomy is often confused with radical separation and complete independence, the autonomy of systems (and, I would argue, human freedom) is strengthened by interconnectedness, elaborate iteration, and feedback—that is, influence. Indeed, these processes create that possibility of change without which there is no freedom. Determinism and autonomy are as inseparable as the multiple aspects of a gestalt drawing.

Thus, Turner, like Kohlberg, recognizes that the social bounds the individual, and vice versa, the individual bounds the social. For the wild "out there" and "within" to thrive in a healthy way it needs to be self-regulated, which is a highly social process that coordinates self and others. In this sense, wildness is at once untethered and bounded, carefree and responsible.

Eisenberg (this volume) highlights a nonhuman form of wild self-regulation when she demonstrates the principle of trophic cascades. For example, left unchecked by an apex predator, "elk grow more abundant and bold, damaging their habit by consuming vegetation," "literally eating themselves out of house and home," and leading to the "loss of biodiversity and ecosystem destabilization." Wild elk need the wolves as much as the wolves need the elk. And healthy—that is, self-regulating—wild nature needs them both. The difference with humans as compared to nonhuman animals, like elk and wolves, is that we're more reflective, and have a much wider range of choices. Our self-regulations therefore need to be greater so as to match our greater capacity for the perversion of wild energy.

4. The Destruction of the Wild

In the field of human-computer interaction, a startling phrase has appeared in the literature. Researchers sometimes now write of their studies being conducted "in the wild." A reader might respond: "Wow, what's going on? Are the researchers, say, taking computers into grizzly country to help them statistically analyze data in the field?" But, no, that's not what the researchers are doing. What they mean by the term in the wild is that the research studies are being conducted outside the controlled laboratory context. The research might be a user study of virtual collaborations (Olson & Teasley, 1996) or how people use cell phones while walking city streets.

We're destroying the wild and hardly know it. We'll continue to read to our children *The Lorax*, and our children will ask us, "But mommy and daddy, who speaks for the trees?" and we'll say, "We all do sweetie"—for we will have reconceptualized trees as spindly growths in a Georgia-Pacific tree farm. A raptor takes off in flight and a second later . . . thud! It crashes to the ground. It forgot that it's shackled in chains by the institution, the zoo, whose mission it is to promote a global conservation ethic. In its advertising campaign, Chevron Corporation shows us a picture of blue herons nesting by an oil refinery, asks, "Do People Care about Nature?" and answers "People Do." The Tundra, Mustang, Outback, Jaguar, Falcon, and Land Rover—the names figment nature and illuminate nothing but our unrecognized longing for the canonical wild.

Future scholarship needs to focus on the mechanisms that allow us to destroy the wild so readily and so quickly, as measured in evolutionary time. Some of the mechanisms are addressed in this volume. The extinction of the wild out there leads to the extinction of the experience of the wild, which sets in motion a downward cycle where we allow for the destruction of more wildness. There is our fear of the wild. There's our hubris in seeking domination over others, which has been reified in many of the religions of the world today. There's our warrior ethos, which sets in motion further domination. There's our need for material goods, which seems insatiable, as corporations know all too well. In Paleolithic times, the environment checked our material desires as readily as it checked our desires for sweets and fats. Now we have to check our desires ourselves, and that is difficult. There's our greed. There is our too-ready acceptance of the faux wild, like chlorinated wave pools that poorly imitate the ocean's edge or jungle safaris in Disneyland. There's our population of over seven billion people and growing, which lies behind virtually every environmental problem, and yet which we evade at most every twist and turn. As we come to understand the mechanisms for how and why we destroy the wild, we'll be in a stronger position to stop it.

5. A Nature Language of the Wild

Some of the examples above highlight that in today's world, we're using the language of the canonical wild for largely nonwild experience—the Tundra and Jaguar name cars, and researchers in human-computer interaction believe they are conducting research in the wild when they simply move outside a controlled laboratory setting. It's important to document how we're destroying the wild through the use of our language. But even more crucial is to recover the language of our experience of the wild, which can then lead us to be motivated to engage and also protect it.

Toward this end, we and colleagues have been involved in an agenda that we call Nature Language (Kahn, Ruckert, & Hasbach, 2012; Kahn, Ruckert, Severson, Reichert, & Fowler, 2010): a way of speaking about patterns of interactions between humans and nature, and the meaningful, deep, and often-joyful feelings that they engender. There are three overarching forms of these patterns: wild, domestic, and perverse. For example, perhaps all of us have had the experience of encountering an animal more wild than not—perhaps a bear, deer, or skunk in the forests, or a large bird that lands close—and we look into that animal's eyes—a powerful experience in itself—and then recognize that the animal is looking into our eyes, and there's that electric moment of mutual recognition, which can last in one's memory for a lifetime. That's a wild form of the pattern we call *recognizing and being recognized by a nonhuman other*. A domestic form of this pattern occurs daily if you have a dog; you both look into one another's eyes, and there's communication. A perverse form happens every day in zoos, though the official signage seeks to stop it. Sometimes zoo visitors will pick up a small pebble or use a piece of food and toss it directly at an animal in a cage. The visitors are trying to get the animal's attention. They want to look into the animal's eyes and be recognized by it. It's a perversion of the wild pattern because the animal is imprisoned and dulled, stripped of its primal autonomy—a vestige of its original self.

We have begun to generate many patterns of a Nature Language, which include, for instance, *sleeping under the night sky, immersing one's body in water, moving away from human settlement, moving one's body vigorously, climbing, cooking around the fire circle, walking the edges of nature, harvesting, being moved by water, sitting by fire, foraging, tracking, killing an animal, bird-watching, and encountering a nature that can hurt us.* Each of these patterns can be instantiated in endlessly different ways and warrants language to speak of them. *Encountering a nature that can hurt us.* It can be lightning during a summer thundershower. It can be the thundershower itself. It can be a rattlesnake or scorpion in the woods; cold or hot weather; ocean waves too big; or sunshine too hot. *Moving one's body vigorously.* In the city we call it exercise, and we might find ourselves inside on a treadmill, with earbuds attached, blasting music to enliven the otherwise-dull experience of running hard and long yet going nowhere. In more wild settings, this pattern can be enacted by running along mountain trails, scampering up a hillside, climbing a tree, swimming across a strong rip current, walking on a ridge, hiking across a boulder field, or strolling through a meadow on a spring day, in tall grasses, with the wildflowers in full bloom and a freshness in the air. Each experience has a story. It has meaning to us, because we're functioning and thriving in relation with wild nature rather than reifying it like a mausoleum.

A full-fledged Nature Language will take some years to develop. As it continues to emerge, it will help lead to the rediscovery of the wild.

References

Kahn, P. H., Jr. (1999). *The human relationship with nature: Development and culture.* Cambridge, MA: MIT Press.

Kahn, P. H., Jr. (2011). *Technological nature: Adaptation and the future of human life.* Cambridge, MA: MIT Press.

Kahn, P. H., Jr., Ruckert, J. H., & Hasbach, P. H. (2012). A nature language. In P. H. Kahn, Jr., & P. H. Hasbach (Eds.), *Ecopsychology: Science, totems, and the technological species* (pp. 55–77). Cambridge, MA: MIT Press.

Kahn, P. H., Jr., Ruckert, J. H., Severson, R. L., Reichert, A. L., & Fowler, E. (2010). A nature language: An agenda to catalog, save, and recover patterns of human-nature interaction. *Ecopsychology*, 2, 59–66.

Kohlberg, L. (1969). Stage and sequence: The cognitive-developmental approach to socialization. In D. A. Goslin (Ed.), *Handbook of socialization theory and research* (pp. 347–480). New York, NY: Rand McNally.

Kohlberg, L. (1984). *Essays in moral development: The psychology of moral development* (Vol. 2). San Francisco, CA: Harper and Row.

Leopold, A. (1970). *A sand county almanac.* New York, NY: Ballantine Books. (Original work published 1949).

Olson, J. S., & Teasley, S. (1996). Groupware in the wild: Lessons learned from a year of virtual collaboration. In *Proceedings of the Conference on Computer Supported Cooperative Work* (pp. 419–427). New York, NY: ACM Press.

Pyle, R. M. (2002). Eden in a vacant lot: Special places, species, and kids in the neighborhood of life. In P. H. Kahn, Jr., & S. R. Kellert (Eds.), *Children and nature: Psychological, sociocultural, and evolutionary investigations* (pp. 305–327). Cambridge, MA: MIT Press.

Rafferty, M. (1970). In H. Hart (Ed.), *Summerhill: For and Against.* New York, NY: Hart.

Turner, J. (1996). *The abstract wild.* Tucson, AZ: University of Arizona Press.

1

Quantifying Wildness: A Scientist's Lessons about Wolves and Wild Nature

Cristina Eisenberg

The wolf is an indicator species. Its tracks signify wildness.

—Field notes, Stanley, Idaho, August 10, 2003

A doe burst out of the forest and tore across the meadow, two wolves in close pursuit. This drama unfolded not twenty feet from where my young daughters and I knelt in our garden peacefully pulling weeds, our pant legs wet with morning dew. One black, the other gray, the black wolf in the lead, they closed in on the doe's haunches. In less than two heartbeats they pierced the woods on the far side of the meadow, leaving a wake of quaking vegetation.

We live at the base of a mountain in northwest Montana, in a lush, verdant valley. The twin mountain ranges that palisade this valley form its real and temporal borders. As wild as the wildest places in the lower forty-eight United States, it isn't quite paradise, although the handful of us who live here think it comes close. Midway up the mountains, overgrown clear-cuts show up as yellow-green rectangles against the darker green of old-growth forest. This area's scant human presence is most conspicuously manifested by those clear-cuts and by the logging trucks that come roaring down the lone two-lane highway. From our cabin, you can walk due east and not encounter much more than wilderness for over one hundred miles, as state forestlands give way to the Bob Marshall Wilderness, a million-acre roadless area.

Landscapes shape and speak to us on a primal level. Most of us have a landscape we intuitively comprehend (Williams, 1994). This is mine. My complex, passionate relationship with this place has inspired me to open my senses to it, and learn to read its nuances and subtleties—to see, hear, touch, and experience it. I open the front door of my cabin and find wolf tracks pressed into the snow. In spring, even before I see the grizzly lumber out of the forest into the meadow to dig roots, I smell its ripe essence. These discoveries give me pleasure and an ineffable awareness of the natural order of things—and wildness.

Humans also have a primal relationship with large predators. This relationship has been eloquently elucidated across the ages in Paleolithic petroglyphs of dire wolves along with other creatures sharp of tooth and claw as well as medieval paintings of wolves menacing humans and sheep. British ecologist Charles Elton (1927) conceptualized the food pyramid, with predators at the top, herbivores in the middle, and vegetation at the bottom. The wolf stands at the pyramid's pinnacle, an apex predator, so called because no other predator controls its abundance or presents a threat to its overall survival as a species. No other predator, that is, besides humans. As a scientist and writer I investigate these relationships—wolf to biotic community and wolf to human—in Rocky Mountain landscapes from Alberta to Colorado.

Science is rooted in empiricism: knowledge based on direct experience. My connection to my homeland has inspired me to use the empirical knowledge and field skills I've acquired to honor its mysteries and embrace its spirit. This combined knowledge—the primordial and scientific—is helping me better understand the predator-prey nexus and nature of wildness. Since moving here in the mid-1990s, we'd been hearing wolves howl from the shoulder of our mountain, but had never seen them—not until that misty August morning when they ran across our meadow.

For some long moments after they passed, my daughters and I knelt motionless in the garden, at a loss for words because of the unprecedented character of what we'd just seen. Then curiosity kicked in, and we stepped outside our small fenced yard to follow the wolves' trail—a trail so fresh I could still smell their musky scent, which held more than a hint of blood and mystery.

I marked one track, and from it we located others laid out in a head-high gallop pattern. As in a tracking 101 lesson, I even found the spot where one of the wolves had turned its head to look at us, because of how this had caused its left front foot to break forward and slightly to the right of its right front foot. But had we not seen the wolves run by, it would've been far more difficult to find, let alone follow, their trail, as most of the tracks were on meadow grass. Too fascinated for fear, we continued to follow the subtleties of their trail—which sometimes consisted of little more than a few bent blades of grass that even as we watched were springing back upright. And I couldn't help but wonder how many other times wolves had run through our land and I'd missed the signs of their passing.

We followed their trail south until it reached the barbed wire fence that bounds our property. The deer had gone over the fence and into our neighbor's land, and the wolves had followed. Rather than jump the fence, though, the wolves had run between two strands of wire, leaving clumps of hair on the barbs. I collected hair, went indoors, and glassed it with a hand lens. The samples had a thick, oily undercoat and long, dark guard hairs—wolf hair. Those wolves had been real enough, and in their passing they urged me to consider questions beyond the pale.

One week later, I sat on a manicured lawn in Middlebury, Vermont, with author Terry Tempest Williams, enjoying the Bread Loaf Writers' Conference. A breeze pushed through the close, humid air, bearing the distant gronking of ravens and green, fertile scent of the hardwood forest that hemmed the hayfield beyond the lawn. Pant legs rolled, barefoot in the grass, we sipped wine and nibbled canapés at the picnic that was an annual tradition begun by Robert Frost. Terry asked about life in Montana, so I told her about having seen wolves hunting on our land. She touched my arm, a look of wonder on her face, and urged me to write about this and tell others how it felt to live with such wildness.

As she said that, I looked out at the landscape beyond the stone fence that bounded the Middlebury campus, to the pale amber of the mown hayfield and brooding green forest that began abruptly beyond that—a forest that had been wolfless for the more than three centuries since the colonists had extirpated all the wolves from the Northeast. I realized then that the knife-edged ecotone between furrowed field and forest held the wolf's story—metaphorically and literally—along with the story of our connection to wildness. It's a story about the darkness of human nature, but ultimately about healing, hope, and social change. Made all the more compelling by our inchoate understanding of wolves and wildness, this story bears telling for the lessons it can teach us about what it means to be human.

Doing this story justice in my scientific research and writing has led me into the gestalt, or unified whole, of the wolf—a whole greater and more powerful than the sum of its parts. In our dark, tangled history with this animal, the real wolf has eluded us, standing in the half-light at the forest edge, mysterious and aloof, known, but in so many ways not yet fully known. Although some see the wolf as one of the most studied species, I argue that we have only just begun to understand the complexity of its behavior and effects on ecosystems.

There is the wolf of science, the wolf of metaphysics, and the wolf of enmity with shepherds and ranchers—three distinct images that elide subtly as they merge in a complex human construct that goes back to the dawn of civilization, and speak to us of our light and shadow sides. Human coevolution with wild animals has shaped our contemporary psyche, creating a longing deep within to re-establish connections we once had with the wild, and dysfunction when these connections become attenuated (Shepard 1978, 1982). As modern humans, however, we are at once drawn to wolves and wildness, and repelled by them. Our quintessential views of wolves have to do with our struggle to understand our place on the earth and the ethical aspects of our existence. They come out of the feeling we have when we gaze up at a shoal of stars: that as much as we know, there are still many things beyond our ken. We fear what we don't fully understand; yet fear of wildness has its place, as we will see.

Wildness is generally accepted as a condition and often defined by what it isn't. For example, with regard to behavior, environmental philosopher and poet Gary Snyder defines wild as that which fiercely resists oppression, confinement, or exploitation. With regard to land, he describes it as a place where "original and potential vegetation and fauna are intact and the landforms are entirely the result of nonhuman forces. Pristine" (Snyder 1990, p. 10). In relation to animals, Snyder defines it as that which is free to live within natural systems. Here I define wild as that which is untamed, and wildness as a condition that allows the full expression of evolutionary relationships between animals and other organisms such as plants. The postmodern perspective suggests that as a species, humans can find redemption if we rewild ourselves and whole ecosystems. Accordingly, rewilding has become a popular trope in conservation (Foreman, 2004). Rewilding involves embracing fear as a primal force that is essential to healthy ecosystems and humans.

To many people today, the wolf symbolizes wildness and freedom from the constraints of city life. As recently as the mid-twentieth century, though, the majority of people in European societies saw the wolf as symbolic of evil and humans' bestial nature (Lopez, 1978). These notions originated in western Europe, fanned by superstition, ignorance, and the Roman Catholic Church, which was determined to repress wildness. By medieval times, humans were obsessed with wolf legends, and widespread werewolf beliefs existed.

Human hatred of wolves enabled this species' ruthless extermination, which reached pathological intensity (Lopez, 1978; McIntyre, 1995). The wolf was extirpated from England between 1485 and 1509, and from Scotland, Ireland, and Denmark around 1770. In other parts of Europe, the wolf's extirpation took longer. The last French wolf was seen in 1934. The last Swedish wolf was killed in 1966, and the last Norse one in 1973, although they have been slowly recolonizing Scandinavia since then (Boitani, 2003; Harting, 1994).

Seventeenth-century colonists brought this fear of wolves and their wildness with them to the New World, and then proceeded to spend the next three centuries systematically and mercilessly exterminating this species in North America (Coleman, 2006). By the 1940s, the wolf had been extirpated from the United States (Young & Goldman, 1944). The influence of European wolf mythology and folklore ran so deep that humans didn't develop a more benevolent attitude until the last half of the twentieth century. This yet-ongoing reversal resulted in the wolf's reintroduction to portions of the United States from which it had been extirpated (Askins, 2002).

As a conservation biologist, I study the ecological effects of wolves on whole ecosystems—and our human relationship with these powerful predators. My quest for the truth about this animal and all it represents ecologically as well as socially has taken me to live periodically with a pack of captive wolves on the Idaho panhandle. It has taken me to the Pierpont Morgan Library in New York City, where I read about human relations with this species during the Dark Ages, and the Aldo Leopold Papers at the University of Wisconsin Archives, where I learned about what it took to open our hearts and minds to wolves. It has taken me to the Adolph Murie and Olaus Murie Papers in Wyoming to learn about how we began to understand the essential role of predation. My inquiry has taken me into wolf dens as a field biologist and to a wolf kill on my land so fresh that the earth was still wet with blood—a killing ground where life and death merged so seamlessly, I couldn't tell where one began and the other ended. It has taken me to contentious meetings with hotheaded ranchers and equally hotheaded wolf advocates—shouting matches so absurd that in retrospect they seem humorous. It has compelled me to spend early morning hours perched atop a high knoll overlooking Yellowstone's Lamar Valley, observing wolf life unfold in all its drama. And it has led me to put in several hundred kilometers of transects in areas frequented by wolves to study their interactions with whole ecosystems. At every turn, the wolves have surprised me—with their courage, with their heedlessness, of the foolishness and politics of humans, with their big hearts and big feet, and with their primal wildness. But far from giving me romanticized notions, my intimacy with wolves has left me with a pragmatic understanding of their feast-or-famine existence and tooth-and-claw role in a biotic community.

My desire to gain greater knowledge of wolves has to do with my ecological interests, of course, but also with my ethical, aesthetic, and

biophilic values. Ethically, I believe I owe other people and future generations the opportunity to share environments with wolves, and gain more
accurate knowledge of what these animals *are*. My studies have deepened
my appreciation of wolves' complex relationships with the biotic community and their extraordinary adaptability in the face of adversity—an
adaptability that has become the sine qua non of survival for species in
our rapidly changing world. My objective as a scientist is to help provide
a more accurate ecological view of wolves, and in doing so, help humans
coexist more peacefully with them. I intend my work to deepen our discourse with wolves so as to learn from them how to live more sustainably
on the earth.

Humans have historically regarded the wolf as a powerful predator
and sociopolitical symbol. Although people have become more tolerant of
wolves, contemporary attitudes range from reverence to hatred. Changing US values about wolves and conservation have enabled the gray wolf's
reintroduction and recovery in fulfillment of the Endangered Species Act,
although an acrimonious debate exists about how to define this species'
recovery and management (US Fish and Wildlife Service et al., 2011). In
the mid-1990s, the gray wolf was reintroduced in the Northern Rocky
Mountain Recovery Area, which comprises portions of Idaho, Montana,
and Wyoming, amid a polarized political climate highly problematic
for wolf conservation. Returning wolves to a landscape dominated by
humans is a challenging enterprise. Human prejudices and sometimes-
thoughtless choices based on economics or expediency have driven other
species to extinction, and nearly did so to the gray wolf in the lower forty-
eight contiguous United States.

Wild animals and places are disappearing from this earth at an unprecedented pace. The wolf is one of the most potent, evocative symbols
of these losses. While extinction is a natural process, in the past century
humans have vastly accelerated it through the rampant and mindless con-
sumption of resources as well as habitat destruction and fragmentation
(Wilson, 2002). Bringing back wolves from the brink of extinction has involved embracing a body politic rooted in empathy and reverence for life
that acknowledges the natural order of predator-prey relations (Leopold,
1949; Newton, 2006; Williams, 1994). Advancing wolf conservation has
necessitated looking hard at our changing relationship with this animal—
and with the wildness its represents.

My research on these relationships has been akin to hiking up one of
my favorite alpine trails in the northern Rocky Mountains. Along the way
I passed through a gradient of habitat and terrain, each offering different

vantage points and insights into wildness, and each offering important survival lessons. When I began doing this work, I feared that by attempting to measure the wild, I would demystify it. But that has not turned out to be the case at all. For even while gathering the most meticulous field data at a wolf den, I feel profoundly touched by the wildness in which I am immersed. This is the paradox of my work: the closer I get to quantifying wildness, the deeper my sense of its mystery.

Part of quantifying wildness involves placing radio-collars on wolves. Why wolves? Because wolves, as apex predators, are proxies for wildness and touch and have the potential to change everything in an ecosystem. And why radio-collars? Because we live in an empirical world, and the data from these collars provide powerful, incontrovertible evidence of the effects of the wildness that wolves represent. Although I use as many noninvasive data collection methods as possible (e.g., track surveys or scat collection), these methods have limitations. To learn more about how nature works, one needs as much data as possible—hard facts, numbers, and the glowing red dots on a Google Earth map that indicate the mark of the wolf's tooth on a landscape. Yet the collars, which can be seen as a violation of an animal's freedom and dignity, are antithetical to wildness itself. So how could it be that the closer I get to quantifying the wildness that wolves embody, the deeper my sense of its mystery?

The answer lies in my relationship with each of the animals I radio-collar. I justify using radio-collars because that is what it takes to obtain data that will help sustain wolves as part of society's attempt to manage wildness. The wolves oblige me by allowing themselves to be trapped and collared. We have a tacit understanding: I handle them with the highest care and respect; they provide a wealth of data that could not be obtained otherwise.

I program the radio-collars to drop off after two years. The first time I picked up a dropped collar, I examined it closely. Its worn black leather strap was embedded with wolf hairs, and it held the musky, clean scent of the young male wolf who had worn it. But beneath that coursed the primal scent of wildness—of swift rivers, jagged mountains, life and death—a wildness so deep I had no words for it. Each time I pore over the data from this wolf's collar and the others displayed on my computer screen, the tidy columns of location coordinates and environmental covariates give me a window into this wildness. These rich data provide astonishing insights about precisely how wolves reshape ecosystems into wholes and reweave the web of life. And this knowing is as essential to our well-being as a species as it is to the wolves (Shepard, 1978; Turner, 1996).

Coming Home to Wildness

I am in Waterton Lakes National Park, in southwestern Alberta. This park lies in the Crown of the Continent Ecosystem, one of only two that contain all species present at the time of the Lewis and Clark Expedition. When I arrived after a winter spent in my Oregon State University lab, saw those rough mountains, and placed my feet on the eskerine earth, my whole heart and every sense expanded, and I felt the sort of bone-deep happiness that is embedded in one's cells. For I had come home to wildness—a wildness more familiar to me than the landscape of my dreams.

Scientific method entails framing hypotheses and testing them via un-biased data collection. It also means patiently looking at things closely. And when one works in a wild landscape and looks closely at ecological relationships, one begins to experience and learn life-changing things.

My wolf research is based on the concept of *trophic cascades*. The term *trophic* refers to anything related to the food web, while the po-etic term *trophic cascades* refers to the movement of energy through the food web when predators are removed (or when they return). This dynamic resembles a waterfall and involves the top-down regulation of an ecosystem, in which predators have a controlling influence on prey abundance and behavior at the next-lower level, and so forth through the food web. Remove an apex predator, such as the wolf, and elk grow more abundant and bold, damaging their habitat by consuming vegeta-tion (termed *herbivory*) unsustainably. Intensive herbivory can lead to deer and elk (termed *ungulates*) literally eating themselves out of house and home, and consequently, to loss of biodiversity and hence ecosystem destabilization. Lacking apex predators, ecosystems can support fewer species, because the trees and shrubs that create habitat for these species have been overbrowsed. With apex predators in them, ecosystems contain richer and more diverse habitat, and thus can support more songbirds and butterflies (Eisenberg, 2010; Terborgh & Estes, 2010). Trophic pat-terns generated by apex predators, furthermore, exert both blatant and subtle influences on animals, including humans, and on the way we see ourselves and our relationship with whole landscapes and other species (Shepard, 1978). Indeed, the reciprocity within the natural world that is the benchmark of trophic cascades is boundless and infinite, and touches all things, even soil and microbes (Eisenberg, 2010).

The concept of *keystone species* is integral to trophic cascades. Ma-rine ecologist Robert Paine (1969) created this term to apply to sea stars, which function as apex predators in the rocky intertidal zone he studied on the Washington coast. There, on a rocky crescent of shore, he hurled sea stars into the ocean. In his control plot he did nothing. As he continued removing stars, the assemblage of species on the rock gradually began to change. Within one year, the sinister implications of his experiment became all too graphically obvious. Where sea stars flourished, so did the vegetation. Where sea stars had been removed, the mussels took over, crowding out other species as well as eating all the vegetation until little more than a dark carpet of mussels and barnacles remained.

Paine's keystone concept is fundamentally about species diversity: the presence or absence of one key species influences the distribution and abundance of a great many others. Research can help us recognize these losses and thereby act to prevent others. This metaphoric term, which refers to species that function similarly to the keystone of an arch (remove the keystone and the arch falls apart), was intended to convey a sense of the unexpected consequences of species removal. Other keystone species besides sea stars include sharks, spiders, and wolves. Processes driven by these ecosystem-regulating species include nutrient cycling and seed dispersal.

Cycling predator-prey populations leave marks on the land. I pay close attention to these marks. This sort of attentiveness inspired classic work on predators and their prey from the 1920s through the 1940s by scientists such as Aldo Leopold and Olaus Murie. In terrestrial ecosystems with large carnivores, one of the most blatant of the marks predators and prey leave is the *recruitment gap*. Wolves create profound changes in an ecosystem not just by killing their prey but also by scaring them. Labeled *the ecology of fear*, these top-down dynamics force elk or deer to nibble plants warily and look up frequently, rather than standing around complacently browsing aspens and shrubs to death. This in turn frees plants from browsing, improves habitat for other species, and increases biodiversity. These dynamics leave powerful imprints. In areas where wolves have been extirpated, for example, no new aspens are able to grow above browse height (about seven feet). Where wolves have returned, aspens are once again able to grow above the reach of hungry elk due to the ecology of fear. In such cases this creates a gap in tree ages, with lots of old aspens, no middle-aged aspens, and lots of young ones.

To document the indirect effects of wolves, I have been measuring elk vigilance behavior (how much time they spend with their heads up scanning for predators versus with their heads down, eating) tree ages and stand dynamics, and songbird biodiversity. I am comparing sites with abundant wolves to those with no wolves to see if a difference exists between them in how trees grow along with the number of songbird species. All of this fundamentally has to do with fear.

Trophic cascades and the ecology of fear are ultimately about relationships—between predators and their prey, or between human and non-human beings. Maintaining and tending to these relationships requires commitment plus vision. As just one of the many scientists currently studying trophic cascades, I am constantly astonished and humbled by the complexity as well as beauty of the food web interactions I observe. And my observations about this numinous web are essentially about the effect of wildness.

The Belly River Wolves

I will tell you a story about wildness. It is simply one of countless tales—some without words, but all stories nonetheless. It is my story, because I admit, as a scientist, wildness and wild places have drawn me in, and are holding me fast, holding my heart and seeking mind captive. It is the wolves' story—the Belly River pack. And it is their ancestors' tale—the hundreds of generations of wolves that preceded them, drawn to this landscape as ineluctably as I am. It is also a story about what scientists know, but cannot say—although that may be changing.

We walked single file on a narrow, overgrown whisper of a trail in Waterton Lakes National Park along the Belly River. This river flows at the feet of Chief Mountain, giving its name to the valley it carved out of limestone—a valley that crosses into Canada and is considered one of the wildest places in this landscape. Young aspens, straight and relatively unbrowsed, the likely product of a trophic cascade, grew closely along either side of the trail, the understory shrubs lush and wet from late spring rains. My group included several field technicians and park conservation biologist Rob Watt, who had worked here for thirty years and knew this place better than anyone. I had been here the previous autumn with Rob. The place he was leading us to was a secret place, its location kept quiet to protect the wolves who denned there.

Quickly we saw evidence that we had traveled in the right direction—wolf scats deposited tellingly every hundred feet or so as bold territorial markers. Occasionally we found enormous, dinner-plate-size grizzly scats. Above the sweet music of riverwater flowing over stones we heard the unmistakable braid of wolfen voices, and around the next bend in the trail heard something big crashing through the woods. This place was not for the fainthearted. Hooting and hollering to let the animals know we were there, we cut upslope. The trail narrowed to nothing at times, but we managed to follow it.

At a slender stream, we found a wolf track the size of a human being's outspread hand pressed deeply into the mud on the bank. Just past the stream, Rob led us off the trail and into the forest understory. A dense thicket of snowberry, serviceberry, rose, and willow soon enfolded us. We passed beaver-cut aspens, their stumps weathered silver. After some heavy bushwhacking, we descended into the swale of a wet meadow ringed by pale-green aspens and inky conifers. And there, on the edge of the meadow, we saw dark hole after dark hole dug into the sloping earth—a series of wolf dens necklacing the south-facing forest edge. We counted six in all, eyes wide with wonder. Yet we were not there to count dens but instead to determine whether there was a pack at all, and if so, whether it had denned and produced pups. This pack, which wildlife managers believed had nine members, had lost seven of them, all killed by ranchers between February and March of that year, when a pack typically prepares to den. Now it was June, and we were uncertain whether enough wolves remained to call it a pack, much less produce pups.

As we walked into the meadow we saw a partially chewed bloody deer skull, with part of its hide still attached, lying in a grassy clearing. For some reason what we were about to walk into didn't register. We didn't think. We simply kept following a wolf trail through some willows. All at once, in a shaft of sunlight in another grassy opening, we stumbled on three small, pale-gray wolf pups playing tug-of-war with a scrap of deer hide. They became aware of us at the same moment we became aware of them and instantly fled into the woods. We stood there, too stunned to say much. We had our answer and left as quickly as possible—our hearts filled with hope.

In August 2007 I returned to locate the active den, testing the hypothesis that areas of most intense wolf use, such as dens, would show the strongest trophic cascades. We bushwhacked into the meadow, clambering over deadfall and unwittingly flushing a chittering flock of winter wrens. We skirted the meadow, examining the dens until we found the one most recently used. Set into the root ball of a Douglas fir, the den was massive and complex—the largest and most finely crafted I'd ever seen—with four openings, each one smooth, deep, and clean, the bare earth finely tattooed with layers of wolf tracks. Clumps of shed wolf hair festooned the wild roses that bracketed the entrance, and pup chew toys—elk jawbones, a beaver skull, and a purloined human shoe—lay strewn nearby.

The Belly River pack had denned here for many years. Some days when I worked at this site, I could feel in my bones that this was its ancestral denning ground and had been the pack's home far longer than we realized. For no matter how often humans shot or poisoned them, these wolves kept returning, their big hearts and big feet leading them along their ancestors' trails, which arrowed toward the dens.

In February 2007, as part of my research we radio-collared one of the Belly River wolves, a yearling male. He wore an Argos satellite collar, which gave me data every two hours. "Please don't be a bonehead," I would admonish him every time I received data on my computer—a red dot vividly marking each location, the dots strung together in a way that indicated he was moving perilously close to ranchlands. On the tenth day after he was collared, a rancher shot him dead. The wolf hadn't been doing anything bad; he hadn't eaten calves or harassed livestock, like some wolves do. But in Alberta, landowners are legally entitled to shoot wolves within five kilometers of their property, for any reason.

Between February and March 2007, ranchers shot six more wolves from the Belly River pack.

In June 2007, we miraculously stumbled on this pack's pups.

In July 2007, a rancher began telling us the story of how he'd trapped and shot another wolf in this pack—an adult female. He'd been able to express milk from the dead she-wolf's nipples, so he claimed to have killed the alpha female. When I heard this story I thought about the pups playing in the sunlight and wondered whether they'd been old enough at the time of their mother's death to survive. Maybe. Another rancher started to relay a story about a "marauding flock" of thirty wolves tearing through his land. Those who knew better shook their heads and laughed at the absurdity of this tale. Others believed it and allowed it to inflame the anti-wolf sentiment that had smoldered deeply in the ranching community for years.

In August 2007, when I returned to the Belly River pack's den, the aspen leaves shimmered gold. I was heartened to find pup scats all over the place—hundreds of them. And one day, as I worked with my field crew to measure the vegetation composition and aspen dynamics around the den, I met a black wolf pup moving in my direction—a different pup than the others we'd seen before. When it saw me, it turned and vanished like smoke into the forest understory.

I checked the den again in June 2008 to see if the pack had produced pups. I found no evidence. We spent that summer searching for the pack, hearing stories on the prairie wind about how this rancher had shot a wolf and that rancher has spotted another from his bedroom window, trotting through his cattle. The immigration officers at the Chief Mountain border crossing occasionally saw the wolves and heard them howling. Nobody had seen any pups.

In June 2009, we revisited the den and found signs that the pack had produced pups again—more pup scats and more chew toys. The Belly River pack persisted, their numbers trailing the vicissitudes of ranching.

I returned in 2010 during the coldest, wettest May on record to see how the pack had fared. It had been snowing off and on all month, with the snow sticking, accumulating, melting, and then accumulating again. We'd had five snowstorms in all thus far that month. On a particularly snowy morning toward the end of the month, six of us set off on a den survey—me, my field crew, and Dan Savage, a veterinarian who had been assisting me with my wolf research. Savage had previously gained years of wildlife veterinary experience working on a wolverine research project in this ecosystem. He was known for his humane work with large carnivores under extreme backcountry conditions (Chadwick, 2010).

Snowflakes the size of silver dollars fell like a heavy lace veil, clumping together, clinging to our cheeks and noses. About eighteen inches of wet snow had accumulated. A line of cougar tracks cleanly pressed into the snow—those of a large male, judging from the tracks' size, width, and roundness—marked the trail. We followed the cougar for a good quarter mile until it left the trail and ascended a small bench. The trail narrowed and the snow intensified, its weight bowing the aspen saplings.

As we had years earlier, we picked up the first set of wolf tracks on crossing the creek—a medium-size wolf traveling in a perfect register trot, its hind feet stepping precisely into the tracks it had made with its front feet— what is called a harmonic gait—the gait wolves use when they are relaxed and moving easily across the landscape to conserve energy. One hundred yards up the trail another wolf joined the first—a smaller animal. A third joined them shortly, this one larger, and a fourth came along from the opposite direction. One of them split off, moving south toward the wolf dens. We stopped and listened, but heard no wolves. I motioned to part of the group to follow me, and the rest stayed behind to minimize our impact.

Three of us proceeded toward the den, moving cautiously along a wolf trail, stopping every few feet to look and listen, communicating in whispers, gestures, and telling glances to avoid disturbing the wolves. We were listening for bark-howls—vocalizations wolves use to communicate distress or alarm. The snow at our feet held the tracks of several wolves coming and going toward the den. The tracks led us over fallen spruces and tangled snowberry bushes, whose unfurling tender-green leaf buds contrasted brightly against the snow. And then I saw the first pup evidence—hundreds of lupine tracks already the size of those typically made by adult coyotes, but much rounder and more robust, scattershot on a patch of muddy ground under the sagging snow-covered boughs of a Douglas fir. Pup scats the size of my little finger, miniature versions of the blatant territorial markers left by their elders, punctuated the welter of tracks.

We stopped again and listened for wolves more closely. Hearing none, we angled in slow motion toward the den, our breath ragged from cold and excitement. We crept forward until we crested the slope that held the den. In an opening near the den's mouth where years earlier we'd seen the pups, we spotted a pup, its tan undercoat showing through its black guard hairs and round rump thickly padded with fat. It scooted across the opening and under a fallen log, possibly joining its siblings. We looked at each other wordlessly. We had our answer, and the sight of that pup felt as miraculous as it had years before. As we carefully backed out, an adult wolf streaked by, its ghostly gray form barely discernible against the snow. It coughed gently toward the pups, signaling our departure. This sequence of events felt peaceful; we'd heard no bark-howls.

We walked a few paces and then had to stop to catch our breaths, somewhat overcome by this experience and the awareness it had given us. More whispers, intimations of wonder, intimations of immortality. We let the awareness sink in that the more wolves came and went and got shot on the prairie, the more they continued to den and leave their progeny and marks on this ecosystem.

After a few moments we continued on our way. And as we walked, I marveled that this place in which I had measured every square inch with instruments—leveling rods, transects tapes, calipers, and clinometers—could still move me to tears, bring me to my knees, and offer me lessons as ancient and primordial as life on this earth. These lessons had to do with wildness, and how profoundly it is a part of who we are as humans as well as how it can continue to inform all we do—every breath, every step, along with our responses to danger, possibility, and the unknown. The lessons related to how rewilding can heal ecosystems—and us.

These wolves' fate has hung in the balance each year that I've studied them, due to our incomplete human acceptance of wildness and poor understanding of how to live with it as we continue human enterprises such as ranching. That there are wolves here at all is because of the delicate rapprochement we have achieved with this species in the decades since we drove it to the brink of extinction. It demonstrates enough reverence for life in some quarters to concede to these wolves the right to exist.

Wolves are proxies for wildness. The answers we are struggling to gain about what wildness means and why we need it lie at the heart of ecological processes such as trophic cascades and the ecology of fear. Trophic cascades are also intrinsic to our ability to live rightly and prosper as a species (Shepard, 1978; Turner, 1996). The Belly River wolves' story—the devastating vicissitudes of their lives as a pack—symbolizes our existential struggle to come to terms with our relationship with the earth and mend our tattered web of life before it falls to pieces.

While these wolves have taught me astonishing lessons about courage, resilience, and hope, they have also taught me that we still have far to go. In the meantime, the wolves keep teaching us, communicating about their powerful effects on ecosystems in ways that are sometimes brazen to the point of being humorous. Always after such lessons I am left reflecting on what scientists know but cannot say: that our relationships with animals are reciprocal, regardless of how empirically we try to look at them. They are not just study units; they are living beings that touch *all* members of the web of life with their presence, including the researchers who study them.

Lessons about the Ecology of Fear

I recently spent a cool September morning with a friend in one of my Montana study sites—a place that has a thriving wolf population. My friend was skeptical about how wolves affect ecosystems. I use the verb *was* because an event that morning rapidly changed his mind.

We visited a large meadow ringed with aspens. The air held a sharp autumn chill and scrim of chimney smoke. A bull elk's ragged bugle cut through the morning stillness. The aspens were at the peak of their turning, blazing like souls on fire. This particular aspen stand had burned in a 2003 wildfire. The blackened bones of downed trees littered the stand, providing blatant evidence of the fire's severity. And among the dead trees stood the legion of straight-trunked aspen that had come up since the fire and had already nearly grown beyond the sapling stage into slender young trees.

Aspens respond favorably to fire; indeed, some ecologists think that this species requires fire to persist and thrive. Fire creates a powerful nutrient surge that sends aspen sprouts shooting up, stimulating a vibrant flush of growth, like the bright thicket surrounding the meadow. Aspens provide essential nutrients for deer and elk, but this relationship can go awry when wolves are absent. Scientific studies conducted in wolfless areas with high elk populations show that burned aspens rarely make it above knee height in these places, because hungry elk mow them down (Baker, Munroe, & Hessl, 1997). Aspens have been declining in the US West since the 1920s, with death from herbivory being a key factor in this decline (Romme, Turner, Wallace, & Walker, 1995). Here in the northern Rockies, however, where wolves abound, aspens grow thick and lush. Many of the aspens in the meadow now reached above browse height.

We walked across the meadow for a closer look at the trees. Along the way I reached down and parted the tawny grass. Ungulate pellets—evidence of deer and elk presence—thickly carpeted the ground, but as we entered the aspen stand, the pellet density dropped sharply. We threaded our way through the aspens, having difficulty walking because the ground was so liberally covered with deadfall and fire debris. The saplings had only been lightly nibbled.

I explained to my friend that the debris made it difficult for ungulates to see predators and thus escape them. Wolves are much more agile in deadfall than their prey. All it takes is one stumble or even one second of hesitation by a pursued deer or elk to give a wolf a lethal advantage. Because of this, where wolves exist, ungulates use aspens, but spend most of their time in the open, where they are safer.

"What about all the rain we've had this year?" countered my friend. "Couldn't that be what's causing all this growth?"

We looked around, seeing the pattern of light browsing repeated throughout the stand. Yet he remained skeptical about the fact that this could be a trophic cascade—a wolf effect. Bottom-up (resource-related) effects, such as the nutrient surge after the fire or three relatively wet years we'd had recently, also contribute to aspen growth. I suggested to him that it was never a bottom-up *or* top-down effect (wolves), because ecosystems are complex. What we were seeing in this meadow was a combination of both of these effects.

As my friend and I stood among the aspens considering the finer points of how this might work, there was an explosion of sound. We turned and saw a white-tailed deer running for its life, legs fully extended like those of a racehorse, tearing downslope about fifty yards behind us. And five yards behind it coursed a lone wolf, closing in fast. The deer hit the aspens and stumbled. We heard a scatter of crashing sounds and then abrupt silence. The wolf had taken down the deer, hidden from our sight by the dense thicket of young aspens.

For some long moments we stood speechless at being vouchsafed a glimpse of a trophic cascade in action. My friend wanted to check out the carcass, but I told him no, that it could mean disturbing the wolf and causing it to abandon its meal. Soon ravens would be homing in on the kill. Grizzly bears and cougars would arrive soon after, drawn by raven calls and the coppery scent of fresh blood. And so we left on that fine autumn morning, talking spiritedly about the eloquent demonstration of the ecology of fear the wolves had provided.

The lessons I have received from the wild animals I study have caused me to re-evaluate my relationship with them as a scientist. I try to be objective, to question and consider all perspectives. Doing so has led to some powerful realizations about our human place in a wild community.

One July dawn my field crew and I went into a large, wet meadow in Glacier National Park, Montana, that contained clumps of scorched aspens, remnants of a 1988 wildfire, and the stands of supple young aspens that had come up afterward. We were there to measure the effects of a trophic cascade on the aspens. This meadow lay in the North Fork Valley, which had a high population of wolves. Mist tendrils wove through the trees, lending an ethereal quality to the landscape.

Before starting work, I scanned for signals from collared wolves, tracing a shoulder-height arc with my telemetry antenna. Nothing. My objective was not to find wolves but rather to avoid them, so as to not disturb them any more than necessary. It may seem paradoxical that someone who collars wolves would avoid encountering them, yet it is one way that I try to minimize the impacts of my research on their community. I only visit sites when wolves may be present when conducting den surveys to determine a pack's conservation status. I put my antenna down that morning and glassed the landscape carefully with binoculars—but still no wolves. I signaled to my crew that it would be fine to carry on.

We proceeded to take meticulous measurements of overstory aspen trunk diameters along with understory shrub species composition and height. We ran our fingers along individual small aspen, feeling for the chisel-shaped wounds left by hungry elk and noting each time a stem had been browsed, engrossed in chronicling the ecology of fear.

Around midmorning the loud drone of an airplane startled us. We could almost make it out overhead through the veil of mist, which was just starting to burn off. I explained to my crew that the plane was likely from the state wildlife agency, there to monitor radio-collared wolf and grizzly bear signals from the air. The airplane made one pass over the meadow, still flying fairly high, and then initiated a second, lower pass. All at once, agitated barking burst from a willow thicket about fifty yards from us. It was a lone wolf, upset about the plane. The wolf continued to bark until the plane left, and then quieted. Most surprising was that the wolf had been there before the plane had come, lying still and hidden from our view, apparently showing no distress at our presence. I considered what to do and decided to resume work, unless the wolf let us know otherwise. We never heard from the wolf again, or saw it, so we continued working until late afternoon.

When we returned the following morning, I explored the thicket where the wolf had lain. I had placed an iron post there two years earlier to mark a songbird point count station—a sitting spot where I had been counting birds to measure the biodiversity in this landscape. I had numbered each post with a bright-blue aluminum disc to make them easier to locate. At these points, I was measuring the indirect effect of wolves on songbird habitat—a painstaking process that required several years of successive observations at each station. When I examined that metallic blue disc now, I found it liberally pocked with wolf tooth marks. Around the metal stake lay multiple wolf scats—adult as well as pup—and a typical assortment of pup toys: old deer bones, plastic water bottles, and a length of frayed fire hose. The wolves has turned my songbird point count station into a rendezvous site—a resting place away from the den where the pack raises their pups. Wolves have one of the strongest bites in the animal world. They can easily bite through an elk femur to suck out the marrow and have been known to bite through steel. I was astounded that they had not destroyed my soft metal disc but instead had only mouthed it gently. Even more remarkable, a wolf had lain there calmly, coexisting with us as we measured the trees, and had only become upset when the airplane buzzed the meadow.

When Leopold's daughter Nina Leopold Bradley (2004) was a college student studying wildlife ecology in the 1940s, her dad used to tell her, "Relationships, remember, it all has to do with relationships." Events such as our meeting with the wolf in the meadow are at once humbling and heartening. I have had many such experiences. They have taught me that as much as I strive to be objective, there is no denying that when studying wild animals, the lessons animals offer often come in the form of the deep generosity and companionship they extend to our kind—even to researchers wielding telemetry antennae and aluminum clipboards who do den surveys.

The Etiquette of the Wild

A couple of summers ago, I was working in another meadow in Glacier National Park measuring trophic cascades in an area that had a high wolf density. Well acquainted with these wolves, I had put collars on two of them and over the years had spent hundreds of days immersed in the landscape in which they lived, measuring their tracks and sign, their trail of carcasses, and how their presence indirectly affected everything in this ecosystem. It was just before dusk. We'd been hearing the wolves howling that day, their voices rising and falling, weaving together, coming from all around us. This paean to wildness felt oddly comforting. Wolves use howling for a variety of purposes, including communication with members of other species (Mech, 1970). We'd also been finding grizzly sign in the willows that grew among the aspens—dinner-plate-size fresh scats filled with roots and shoots. Many bears and other carnivores roam the area where I do my research; in fact, some say it contains one of the highest densities of large carnivores in the United States (Kunkel, Pletscher, Boyd, Ream, & Fairchild, 2004; Weaver, 2001). To maintain peaceful relations with these creatures, I avoid their foraging grounds during crepuscular periods—dusk and dawn.

We had one more plot to measure to complete our survey, but it was late, and I sensed it would be best to leave. We marked it, six feet in radius and on the edge of the aspens, with pink pin flags arranged in the cardinal directions. As we filed out of the meadow on a well-worn game trail, the wolf howls continued. Piles of elk scat marked the trail, along with abundant wolf, cougar, and bear scat, supplying bold proof that this was a heavily used carnivore trail as well.

The following morning we returned to finish our survey. As we approached my plot, there in its center lay a carcass—or to be more precise, one-quarter of a freshly killed elk. A scatter of bloody wolf tracks lay around the hunk of elk like scarlet blossoms on grass. We looked at the meat incredulously. It had not been there the previous day, but here it was—most of the hindquarters of an elk, with the backstrap still attached, all neatly contained within my plot. As we pondered the meaning of all this, we heard a guttural huff from just inside the aspen stand. There, backlit by the morning sun, stood a grizzly bear, rearing up on its hind legs, announcing its keen mutual interest in the elk meat. I had prepared my crew for such situations, when showing fear would only provoke an attack. We left, walking calmly and straight backed out on the carnivore trail that had taken us into the meadow.

One week later we returned to finish the plot, but met the same bear, napping just inside the aspens this time; the bear huffed at us again. The elk carcass had been reduced to little more than sinew and bone by then, but the bear was still possessively positioned, perhaps hoping for another gift from the wolves. We left again. It wasn't until two weeks later that we finally were able to finish surveying my plot. In it I found many young aspens, growing straight and true, and only lightly browsed, offering abundant evidence of a trophic cascade.

I never did figure out whether the meat was an accident of fate or a gift. But this experience taught me much about our connections to the wild—how if we honor the howl of wolfen voices, the sleeping bear, and the primacy of a freshly killed elk along with the gift its meat provides to whole ecosystems, we will be that much closer to learning to live sustainably on this earth. And it made me reflect on who we really are beneath the layers civilization has imposed on us.

The Stoney Flats Grizzly

Coexisting with bears as I do while engaging in wolf research has supplied some of my deepest lessons about wildness. Hundreds of inadvertent meetings with grizzly bears in my field sites have taught me that only when we open ourselves so completely to wildness that we lose our modern sense of ego, and become humble and open, are we vouchsafed passage through life-threatening situations. Wildness is simply wildness. Beyond the romantic ideal depicted by Henry David Thoreau (2004) and John Muir (2010), true wildness is uncompromisingly and unforgivingly blatantly about survival on a primal level. Make one misstep and you can easily end up as food for another animal (Jans, 2005; Krakauer, 2007). This close-to-the-bone edge has drawn contemporary iconoclasts and philosophers Edward Abbey (1990), Gary Snyder (1990), Doug Peacock (1990), and others to wildness. They have found that contact with wildness can bring about redemption and healing, because it forces us to realign ourselves with values that kept our species whole and functioning millennia ago. This can be at once the most grounding and frightening aspect of wildness.

Picture this: you and four companions are hiking along a Rocky Mountain trail, with a dense aspen forest on either side. You are in the lead on this preternaturally beautiful late summer day, and your senses are wide open. Thistledown floats on the breeze and flocks of neon-bright bluebirds preparing to migrate dart about, gleaning the season's last insects.

It's the kind of day that reminds you of your own mortality. As you walk, you talk about food and life, while singing and laughing, feeling so glad to be alive. You had heard an aggressive grizzly bear was prowling the area, and when you first stepped on the trail, had gotten a strange feeling in your gut. As you walk, you keep looking for the bear, but don't see it. Halfway down the trail, though, hunkered down in the shrubs at the forest edge, the bear waits.

Bears are considered a "generalist" species, which means that a wide range of habitat and food meets their needs. While they tend to be patient with humans, food issues can put humans in conflict with bears. This is exactly what happened two months down the line from the series of seven late-spring snowstorms in Waterton in 2010—a force of nature, right? Well, yes, and no. While the unseasonable snow was undoubtedly a force of nature, this disturbance had an anthropogenic aspect. Human use of natural resources—timber harvest and fossil fuel combustion—have increased atmospheric carbon dioxide, causing global heat to rise and weather disruptions. The Waterton storms had delayed the ripening of all the berries by over one month. While the grizzlies had managed well in the interim by eating roots, shoots, and insects, they had become distinctly out of sorts on discovering that their favorite berries were unavailable when they wanted them.

I went to Waterton in late August of that year to measure the aspens in a fire area in the Stoney Flats, which is a large prairie patch bounded by an aspen parkland to the south and a robin's egg blue, glacier-fed lake to the north—a prime elk winter range. Two weeks before my arrival, a grizzly had bluff charged a friend photographing mountain goats in one of my study sites, coming heart-stoppingly short of physical contact. That summer there had been a total of six bluff-charge incidents in the park. One had been videotaped (shaky handheld footage shot by a park visitor; what was he thinking videotaping an aggressive bear?) and had taken place in the Stoney Flats. It involved a healthy, three-year-old amber-colored male bear with chocolate-brown face and legs.

The bear situation was bad elsewhere. In Yellowstone National Park, earlier that summer a boar had killed a researcher, and a sow had killed and eaten a camper. I hoped that now that the berries were finally ripe, the bears would be so busy eating them that they would leave us alone

Stoney Flats lay three miles up the Wishbone Trail. Bears favored this flat, shoestring route hedged by aspens and berry-laden shrubs, as evidenced by abundant bear tracks and dinner-plate-size purple scats. I had done two years of songbird surveys here, which called for walking this trail in the predawn darkness, my path illuminated by a weak headlamp beam as bruins peacefully foraged on either side.

The Wishbone Trail had been closed for six weeks because of the grizzly in the video who had harassed other hikers and researchers. But given the time-sensitive nature of my work and my bear experience, park administrators and I had decided I should carry on. Before starting work I had briefed my crew about bear safety: make plenty of noise to let bears know you are there; stay calm but don't act like a victim; and talk to the bear firmly yet respectfully. A predatory bear is as likely to come from the rear as from the front. Screaming and running will provoke a mauling. And above all, remember that stepping aside from a bear gracefully is not an act of submission but rather an act of admission that other members of a food web take precedence when we meet them on their own terms.

My crew consisted of five persons. I had configured us in a "sandwich" formation: those with the most bear experience in the front and rear, and everyone else in the middle, where they would be best protected. We proceeded to put transects through the aspens, to measure wolf effects in this landscape. The trees were in splendid health, showing fine proof of growing above the hungry mouths of elk. Everything was going well, and our data was coming in true, until the end of the fifth day.

We were leaving on the Wishbone Trail, moving briskly and talking loudly. I had been scanning for the Stoney Flats grizzly, every sense honed, because of an odd feeling I had that he was there, waiting. And then about one mile from the end of the trail (and my car), we heard a heavy thud on the beaten earth behind us. It was him. In fact, I had just walked right by him, so close that had I reached out, I would have touched him where he lay hidden in the vegetation alongside the trail, waiting for us.

The grizzly began to follow us, about twenty feet away. I kept our group moving. We looked over our shoulders as he approached, head low, hackles up, jaws clacking, hyperventilating and salivating, dark eyes like coals smoldering into ours. In an even voice I asked everyone to take out their bear spray. The grizzly was beautiful, well-muscled, and plump from gorging on serviceberries. And the moment my eyes met his, I understood what he wanted to do to us.

He closed the gap to ten feet. We kept moving, not breaking our pace, and began to scold him like an unruly child. After about three minutes of this, the bear sat on his rump sideways on the trail, hind legs splayed in front of him and front paws between his legs. He cocked his head and looked at us, with a forlorn, utterly confused expression on his face. Clearly he wasn't used to humans reacting to him in this way, so he stayed put. We continued on our way, and I radioed park dispatch. The wardens soon arrived, but none of us saw the bear again that day, or when we returned to finish our work three days later.

Climate change driven by human overconsumption of resources helped trigger the late-season snowstorms. The resulting disruption to the grizzly bears' food supply caused them to take out their frustrations on humans. This provides a harrowing lesson as we move into a future that includes a burgeoning human population, escalating use of fossil fuels, and accelerating global change—along with few solutions to these issues.

The day after we met the Stoney Flats grizzly, I visited the Belly River pack's den to survey it for wolf use. While a field technician and I sat at the den's mouth, a large bear calmly foraged in the willows fifty yards away. Butterflies flitted through the air. We talked about bears, wolves, and wildness, and how coexistence comes naturally—it's in our genes, part of what makes us human. And if we open ourselves to the lessons wild nature has to offer, via powerful messengers such as the Stoney Flats grizzly, perhaps we can relearn how to deal with fear and wildness. Perhaps we can even learn how to mend the damage we've done to the earth.

Three months later, as the snow lay deep and the solstice neared, I reflected on this and the other lessons about wildness I've received. The animals I work with—the wolves and bears I inadvertently come into contact with while doing my research—have been powerful teachers for me. I continue to study wolves and trophic cascades in ecosystems throughout the West. Quantifying wildness in this manner is integral to reweaving the web of life. But more than this, such awareness is essential for conserving the human spirit and rewilding ourselves.

References

Abbey, E. (1990). *Desert solitaire: A season in the wilderness*. New York, NY: Touchstone.

Askins, R. (2002). *Shadow Mountain: A memoir of wolves, a woman, and the wild*. New York, NY: Doubleday.

Baker, W. L., Munroe, J. A., & Hessl, A. E. (1997). The effect of elk on aspen in the winter range in Rocky Mountain National Park. *Ecography, 20*, 155–165.

Boitani, L. (2003). Wolf conservation and recovery. In L. D. Mech & L. Boitani (Eds.), *Wolves: Behavior, ecology, and conservation* (pp. 317–340). Chicago, IL: University of Chicago Press.

Bradley, N. L. (2004). Interview by Cristina Eisenberg, September 29, Baraboo, WI.

Chadwick, D. (2010). *The wolverine way*. San Francisco, CA: Patagonia, Inc.

Coleman, J. T. (2006). *Vicious: Wolves and men in America*. New Haven, CT: Yale University Press.

Eisenberg, C. (2010). *The wolf's tooth: Keystone predators, trophic cascades, and biodiversity*. Washington, DC: Island Press.

Elton, C. (1927). *Animal ecology*. Chicago, IL: University of Chicago Press.

Foreman, D. (2004). *Rewilding North America: A vision for conservation in the 21st century*. Washington, DC: Island Press.

Harting, J. E. (1994). *A short history of the wolf in Britain*. England: Pryor Publications.

Jans, N. (2005). *Grizzly maze: Timothy Treadwell's fatal obsession with Alaskan bears*. New York, NY: Penguin.

Krakauer, J. (2007). *Into the wild*. New York, NY: Anchor.

Kunkel, K. E., Pletscher, D. H., Boyd, D. K., Ream, R. R., & Fairchild, M. W. (2004). Factors correlated with foraging behavior of wolves in and near Glacier National Park, Montana. *Journal of Wildlife Management, 68*, 167–185.

Leopold, A. (1949). *A sand county almanac: And sketches here and there*. New York, NY: Oxford University Press.

Lopez, B. H. (1978). *Of wolves and men*. New York, NY: Touchstone Books.

McIntyre, R. (1995). *The war against the wolf*. Stillwater, MN: Voyageur Press.

Mech, L. D. (1970). *The wolf*. Minneapolis, MN: University of Minnesota Press.

Muir, J. (2010). *My first summer in the Sierra*. New York, NY: Echo Library.

Newton, J. (2006). *Leopold's odyssey*. Washington, DC: Island Press.

Paine, R. T. (1969). A note on trophic complexity and community stability. *American Naturalist, 103*, 91–93.

Peacock, D. (1996). *Grizzly years: In search of the American Wilderness*. New York, NY: Holt.

Romme, W. H., Turner, M. G., Wallace, L. L., & Walker, J. S. (1995). Aspen, elk, and fire in the northern range of Yellowstone National Park. *Ecology, 76*, 2097–2106.

Shepard, P. (1978). *Thinking animals: Animals and the development of human intelligence*. Athens, GA: University of Georgia Press.

Shepard, P. (1982). *Nature and madness*. Athens, GA: University of Georgia Press.

Snyder, G. (1990). *The practice of the wild*. New York, NY: North Point Press.

Terborgh, J., & Estes, J. (Eds.). (2010). *Trophic cascades: Predators, prey, and the changing dynamics of nature*. Washington, DC: Island Press.

Thoreau, H. D. (2004). *Walden*. New York, NY: Beacon Press.

Turner, J. (1996). *Abstract wild*. Tucson, AZ: University of Arizona Press.

US Fish and Wildlife Service, Montana Fish, Wildlife, and Parks, Nez Perce Tribe, National Park Service, Blackfeet Nation, Confederated Salish and Kootenai

Tribes, Wind River Tribes, Washington Department of Wildlife, Oregon Department of Wildlife, Utah Department of Natural Resources, & USDA Wildlife Services. (2011). C. A. Sime & E. E. Bangs (Eds.), *Rocky Mountain wolf recovery 2010 interagency annual report*, Helena, MT: US Fish and Wildlife Service, Ecological Services.

Weaver, J. L. (2001). The transboundary Flathead: A critical landscape for carnivores in the Rocky Mountains. WCS Working Papers No. 18, July. Retrieved from http:www.wcs.org/science.

Williams, T. T. (1994). *An unspoken hunger*. New York, NY: Pantheon.

Wilson, E. O. (2002). *The future of life*. New York, NY: Vintage Books.

Young, S. P., & Goldman, E. A. (1944). *The wolves of North America* (Vols. 1 & 2). New York, NY: Dover Publications.

2

The Wild and the Self

Jack Turner

In spring 1989, while returning from a walking meditation retreat in Death Valley, our little group stopped at a Mexican restaurant in Lone Pine, California. We scattered to several tables, and—rather pathetically hungry for news—grabbed various pages of an abandoned newspaper. We ordered courses and casually glanced through the usual unnerving trivia. Then someone said, "Edward Abbey died."

No one knew what to say; the article provided little detail. But I felt something important had ended, as when the loss of a warrior in battle seems a prelude to defeat.

After we returned home to our zendo, I walked up the hill to Gary Snyder's house. I thought that he might not have heard the news and would want to know. When I entered, Snyder was sitting with Jack Loeffler (2003), one of Abbey's best friends and the author of *Adventures with Ed: A Portrait of Abbey*. Loeffler (1989) was visiting Snyder to interview him for his collection *Headed Upstream: Interviews with Iconoclasts*. I couldn't ask for better company.

They knew. The mood was somber.

Snyder poured me a glass of nearly frozen Herradura tequila, and we sat around the table telling stories. Loeffler reminisced about Abbey and their rich life in the desert; Snyder talked about Japan, Zen, poetry, and the many other things in which he is so fluent; and I recounted my travels in the Himalayas. Late at night, in pitch darkness, Snyder took me to the path that led to my tent in the meadow below the zendo. He gave me a sheaf of papers and said, "You really ought to start writing those stories down."

When light came, and I sobered up, I looked at the papers. One was "The Etiquette of Freedom," an essay eventually published in *The Practice of the Wild* by Snyder (1990). I read it several times that morning while lying in my sleeping bag among the shimmering Ponderosa pines.

I pondered Snyder's compliment the night before. That, I realized, would probably launch my writing career, and his essay would furnish my subject.

"The Etiquette of Freedom" is the best essay ever written about wildness, Roderick Nash's *Wilderness and the American Mind* (first published in 1967) is the best book on the subject, and Abbey's many-faceted works explore the varieties of wild experience. They are the closest things we have to a modern canon on wildness. I am always alert for distinguished additions, however.

Bill McKibben (2008) offers a fine collection of essays, poems, and songs—by Joni Mitchell and Marvin Gaye—essential to understanding American nature writing in his recent book *American Earth: Environmental Writing since Thoreau*. I systematically read everything McKibben writes, and was looking forward to *American Earth*. The book's thousand pages are both impressive and oddly lacking. There is only one essay by Abbey, no excerpt from Nash's *Wilderness and the American Mind* (or any of his many essays), no essays from Snyder (though two fine examples of his poems), and not a whiff of bioregionalism, deep ecology, or Earth First! It's high-minded, liberal, and unsullied. There's no mucking about in the swamps of contention here; everyone marches to the same drummer.

I found McKibben's comment on the wild in his introduction to *American Earth* to be disheartening:

Of all the images and the metaphors that have fueled environmentalism in the past, and that have been strong enough to accomplish these good and necessary works, the most important has been that of the wild. The wild, the pure, the clean: from Thoreau to Abbey, from Muir to Austin to Brower, that's been the key idea, the emotional trigger. It's been fuel enough to take us to the orbit that we needed to reach in the past. But as we set about the work that faces us now—the work of reorienting our lives to ward off the apocalypse that science now predicts—we must continue to find further images, further metaphors. (*American Earth*, p. xxv)

What's wrong here?

Well, the wild is basically neither an image nor a metaphor, although it could be either. It is a descriptive term denoting a *quality*, which, along with its historical development, receives an extensive entry in the *Oxford English Dictionary* (OED). Like all descriptive terms it can generate facts. "It's a wild orchid, not a cultivated orchid" can be as much a fact as "It's a red car, not a blue car."

I feared that the phrase "images and metaphors" would marginalize the wild in the physical sciences, or abandon it as merely figurative, even though I like to assume that what we are all talking about is *wild nature*. For all his praise of the concept, McKibben supplies a conclusion that seemed bland. Compare it to the microbiologists Lynn Margulis and Dorian Sagan (1995, p. 55): "Life as God and music and carbon and energy is a whirling nexus of growing, fusing, and dying beings. It is matter gone wild . . . " That's the spirit we need to capture!

The audience for wildness has scattered into a plethora of other legitimate causes—not only the disaster of massive climate change, but also the difficulties of establishing women's rights, the horror of Tibet, ongoing population explosion, quality of our food, ad infinitum. At the same time, the cause of conservation has fractured into legitimate concerns—specialized political niches for saving the redwoods, polar bears, whales, corridors, diversity of nearly every place on the earth, ad infinitum. The bookracks are sagging with doom.

Instead of seeing wildness as a doddering concept on the brink of extinction, I propose we embrace it anew, grounding our efforts in the rich vein of US usage and texts, the physical sciences, and an infusion of Taoist ontology. I propose amplification and refinement, not abandonment, a top-down appreciation, reverence, and even love for the whole thing, all of wild nature, from the chemical elements we inherited from exploding stars, and that still course through our veins, to Gaia, the functioning of our immune system, and the stalking bacteriophage. I propose this for the whole thing—*wild nature*—from the trillions of stars down to the trillions of synapses—the whole damn thing.

I further suggest we associate the concept of the wild with the concept of self. I don't mean the psychological self, the bearer of our personal identity, the focus of our agency, or the self we often speak of as like a homunculus located somewhere between our brow and big toe, analogous to a penny in our pocket. No, I want to consider the self of a singular being, the self of self-replicating DNA, the self of autonomous systems, and the self of the term *tzu-jan* (*ziran* in the Pinyin system of romanization), a Chinese compound that literally translates as "self-so," the spontaneous newness that pours forth each moment, in the world and our experience, and has given us the gift of it all.

I will ignore here our experience of the wild, our phenomenological descriptions of it, and its obvious association with freedom. I have a lot to say about those matters, but they deserve their own essay—or book. My task here is to elaborate wildness in its varied forms, for it now seems encapsulated in a narrow conservation discourse that inhibits its conceptual enrichment.

Nature (the old capitalized substantive) and wildness are not always marked by differences in the language of other cultures, and attempts to translate them frequently result in confusion, or worse, gibberish—rather like trying to translate *God* into classical Chinese: there is no God there, despite the conniptions of early missionary translators. There is no substantive *nature* in the Hebrew Bible; there was none in Japanese before its cultural contact with China. That has not inhibited the impact of the Bible or early Japanese culture on world history.

Some may dread an elaboration of our beloved wildness on the grounds that it reeks of cultural imperialism. I'm not worried. Cultures accentuate aspects of collective experience and elaborate their fundamentals. For example, the classical Greeks went beyond nakedness to elaborate what Kenneth Clark (1972), in *The Nude*, called the "ideal form" of the human body. For 2,500 years, representations of the nude have been a recurrent theme in our drawing, painting, and sculpture, from Praxiteles to the immensely varied modern nudes of, say, Pablo Picasso, Amedeo Modigliani, Lucian Freud, and Ralf Gibson. It is a vein we are still actively mining.

Not so in Chinese culture. In his remarkable book *The Impossible Nude*, the French sinologist François Jullien (2007, p. 47) says of the nude:

China simply *missed* it. This brings up the questions, what prevented the development of the nude in China, and what other possibility prevailed to the point of shrouding the nude and blocking it off entirely? Or more radically: what, in cultural terms, is a possibility that fails to develop, and how should it be understood? To me there is nothing anecdotal about this question, which cannot even be classified with the vast but nowadays so clearly mapped out field of anthropological research. The substance of the question is philosophical, and this substance is what I want to bring to light and examine from the starting point of China and the nude.

Or consider a positive example: the appreciation and elaboration of tea, and its use in Japanese culture. Wild tea is native to a geographic juncture of Assam, Burma, China, and Tibet; it has a long history of cultivation in China, and was associated with Buddhism. Yet tea along with its utensils, serving, aesthetic, and philosophy all achieved a subsequent elaboration as well as refinement in Japanese culture that is now world famous—an aesthetic with profound consequences for the history of that nation.

We've *missed* a lot of things too—and failed to elaborate others. Let us not fail to elaborate a rich appreciation of wildness. Cultures elaborate at will, and each elaboration is contingent, each fascinating in its particularity. There is nothing insidious or malicious about an American elaboration of wildness. So . . . let's do it.

I offer several immodest proposals, or an invitation to think anew about a concept we increasingly take for granted, in order to fertilize and water what appears dormant or perhaps even moribund. I assume my elaborations will be constructive. My main sources are texts, the usual materials for conceptual inquiry, an abstract path fraught with peril, and a path that's easy to get lost in. It has happened to me.

My emphasis on the self as a singular is at odds with our current stress on what Barry Commoner in *The Closing Circle* (1971) called "the first law of ecology": that everything is connected to everything else—a general principle, if you will, espoused by seemingly everyone from Hua-yen Buddhists to Greenpeace. But what about that *thing* in *everything*, the *parts* of those *multiple interconnected parts*, to use Commoner's language, the stuff of nature that endures for at least a moment, and is composed of waves and particles, so we're told, that can have properties, that is present in processes and events. What about those things? I'm not thinking of anything special here—just snails, quarks, molecules, aspen trees, clouds, tribes, ecosystems, and the dwarf galaxy in Sagittarius. Although criteria for individuality vary, each creates a "thing" that is a part of nested hierarchies, any or all of which may be wild. We have only to consider ourselves. The human organism consists of multitudes of simultaneously interacting particulars—cells, tissue, bones, hormones, organs, pheromones, and dendrites—trillions of things, all interrelated and connected, and all singular. How are these particular, singular selves related to wildness?

I begin with the word (concept) wild. It is a wildly popular word, at least in the United States. We have a slew of movies (*The Wild One*; *The Wild Bunch*; *Into the Wild*; *The Wild Child*; *The Wild, Wild America*; *Wild Hogs*; *The River Wild*; *Wild at Heart*; *Wild Bill*; and *The Wild Blue Yonder*), and shelves of books (*Into the Wild*, again; *Where the Wild Things Are*, again; *Wild: An Elemental Journey*; *Wild at Heart: Discovering the Secrets of a Man's Soul*; *Wild Justice*; *The Call of the Wild*; *The Wild Trees: A Story of Passion and Daring*; *The Wild Birds: Six Stories of the Port William Membership*; *Wild Card (Elite Ops)*; *Wild Fermentation*; *Across a Wild and Dark Sea*; *The Practice of the Wild*; *The Abstract Wild*; and a plethora of mass-market romance novels with wild in the title). We have songs ("Born to be Wild," "Walk on the Wild Side," and "Into the Wild Blue Yonder"), television shows (*Wild Kingdom* and *Pretty Wild*), innumerable videos (*Girls Gone Wild*), a sports team (Minnesota Wild), a foundation (Wild Foundation), a *New York Times* blog (*The Wild Side*), and catalogs of wild adventures and wild animal parks.

We have charming oddities like Wild Diva Shoes. And we have poems, of course, many poems, one lovely one, for instance, by Emily Dickinson (1890/1961, p. 32):

Wild nights! Wild nights!
Were I with thee,
Wild nights should be
Our luxury!

Futile the winds
To a heart in port,
Done with the compass,
Done with the chart.

Rowing in Eden!
Ah! the sea!
Might I but moor
To-night in thee!

Viewing, reading, and listening to this, one would conclude we are intimate with the concept of the wild and that the encapsulated environmental use of the word reveals a quaintly narrow perspective. And note: popular usage is often close to the truth. We would do well to recall the prescription offered by Ludwig Wittgenstein (1953, section 43): "For a *large* class of cases—though not for all—in which we employ the word 'meaning' it can be defined thus: the meaning of a word is its use in the language." Our usage of the term wild is broad and rich.

That goes for the historical development of the word too—one that provides a salutary introduction to the concept. The *OED* lends its authority to the breadth (or more formally, extension) of the word. It is probably an old word, much older than dictionaries. The *OED* traces its history to the *Corpus Glossary*, a Latin/Anglo-Saxon dictionary compiled in AD 725. Interestingly, it gives the Latin equivalent *indomitus*, a word that suggests much about the subsequent use of wild.

The first entries are commonplace: "1. Of an animal: Living in a state of nature; not tame, not domesticated." There is an important difference between the words tame and domesticated. Tame implies broken, such as when a wild horse is broken. American Indians occasionally tamed bears, but never domesticated them. Domestication implies habituation to human society and artificial selection (not natural selection) according to cultural, economic, or aesthetic criteria. (It has been suggested that the difference between the domesticated—such as dogs, cattle, or pigs—and the undomesticated may be that captivity renders some species sterile, thus assuring their continued wildness.) Then too, many species have never been either tamed or domesticated—Anna's hummingbirds, western diamondback rattlesnakes, and great white sharks, for example. And what to do about cloned cows and farmed salmon? Language falters. Are they organic machines?

"2. Of a plant (or flower): Growing in a state of nature; not cultivated." The cultivation of previously wild species has been commonplace for 10,000 years (at least); now we have genetically engineered species. Are they tamed or domesticated? Language goes on holiday. The ancient and useful dichotomy of wild/cultivated will require expansion as genetically modified crops become prevalent, but in any case, the artificial expands even as the wild declines. I find it disheartening that E. O. Wilson (2006) now supports genetically engineered crops. Even more worrisome, we now have disciplines—synthetic and digital biology—that seem to demean "natural" biology and cherish the artificial. Indeed, some see the digital universe as analogous to a biological system of autonomous, self-replicating codes.[1]

As Snyder (1990) said in "The Etiquette of Freedom," the dictionary is only noting what wildness is not, and that suggests we urgently require an elaborated analysis of wildness and its benefits. The phrase *state of nature* is not particularly helpful either, as though something wild could perhaps not be in a state of nature.

"3. Produced or yielded by wild animals or plants; produced naturally without cultivation; having the characteristic (usually inferior) of such productions . . . wild silk." One wonders where the usually inferior comes from and what the argument might be for that claim. Has our taste become so palsied that we prefer fillets of insipid salmon to the tangy original? Is the *OED* prejudiced against the wild?

"4. Of a place or region: uncultivated or uninhabited; hence waste, desert, desolate." More negatives and a further indication that we must consider the cultural prejudices of dictionaries. England is not noted for its deserts, but if the *OED* editors spent several days wandering the Sonoran Desert with conservation scientist Gary Nabhan, they might revise their definition in appreciation of deserts, removing the notion of desolation altogether. The flora and fauna along with the mountains and arroyos of the Sonoran Desert are not a waste. And is there something wrong with being uncultivated or uninhabited?

Then things get more interesting. "5. Of persons (or their attributes): uncivilized, savage, uncultured, rude, not accepting, or resisting, the constituted government; rebellious." The *OED* offers as examples "wild Irish" and "wild Scottes." Imperialism rears its ugly head. I think of *Civil Disobedience* by Henry David Thoreau (1849/1973) and its original, more bracing title, *Resistance to Civil Government*. I would like to ask Thoreau what he thinks of *A Year of Living Dangerously*, *Blood Diamonds*, and *Kite Runner*. Having been in places with no civil government, or where it has collapsed, I admit to a fondness for them. I've felt the tension between the wild and the civil. But then Thoreau was also ambivalent about the wild—a tonic, yes, but as a life. . . . He didn't think so, and I don't either. Let's admit to ambivalence; it is a bit too facile to go about saying that all good things are wild and free. After all, a capacity to tolerate ambiguity and ambivalence is a sign of maturity.

At this point an American dictionary undoubtedly would list wild Indians—Crazy Horse, Sitting Bull, and Geronimo—individuals and their tribes now usually deprived of historical context and hence valorized. *Blood Meridian* by Cormac McCarthy (1985), however, suggests that living in Indian country might not have been much fun for outsiders. Today, consider living along the border of Pakistan and Afghanistan (an arbitrary line drawn by a forgotten British civil servant) among the Pashtuns, who still maintain tribal territories, and over whom the Pakistan government enjoys not one smidgen of control. Truly wild people, by our standards, but a people who have a functioning order thousands of years old, and for whom the rest of Pakistanis, and North Americans, are outsiders. The term wild now marks a set of social distinctions: wild/uncivilized, inside/outside, or the other/us.

The contrast between uncontrolled and controlled deepens.

"6. Not under, or not submitting to control or restraint; taking, or disposed to take, one's own way; uncontrolled. Primarily of animals and hence of persons and things along with various shades of meaning. Acting or moving freely without restraint; going at one's own will; unconfined, unrestricted."

The negatives continue, but new terms appear—terms that imply philosophical issues, such as *free* and *will*, or combined, the philosophical thicket of *freewill*. *Indomitus* versus control, the heart of the matter, the hinge on which turns the defense of the wild against its many enemies.

For example, there is Fredrick Winslow Taylor's scientific management, a method of controlling workers that has developed into a general ideology of control, whether over people, lands, species, or forests. From the early Aldo Leopold to contemporary conservation biologists, this ideology has been embraced. The increasing management of nature —control for its own good, of course (not to mention others' profit)—has become *normalized*, in Michel Foucault's sense.

An insistence on wildness instead of control fractures the foundations of modern conservation. Wildlife management becomes an oxymoron. After all, the state claims ownership of grizzly bears (or more accurately, has placed them under its police powers and thus control). Does anyone at this point in history believe wildlife in the United States is wild and free? And yet . . . how about that 600-pound boar you just ran into on the trail? He's not wild? We are obviously dealing with a word whose meanings are multiple, complex, and sometimes contending.

And deeper still: "7. Not submitting to moral control; taking one's own way in defiance of moral obligation or authority; unruly, insubordinate; wayward, self-willed." Wildness also marginalizes morality. Resistance to authority, of all varieties, is privileged. One is reminded of the anarchist pedigree of the wild bunch, from Thoreau to James Dean to Abbey, T. E. Lawrence and the Arab Revolt, or Chaung-tzu's rejection of conventional values. *Taking one's own way*. Thoreau's passage in *Walden* sounds the *obligato*: "If a man does not keep pace with his companions, perhaps it is because he hears a different drummer. Let him step to the music which he hears, however measured or far away" (Thoreau, 1854/1971, p. 326). Are we willing to let others act according to these principles, if they can be described as principles? More ambivalence? Most important for my discussion, this seventh constellation of meanings introduces the term self-willed—the key to my elaborations of wildness.

The next entries seem increasingly negative in implication and remind us, again, of the ambivalence at the root of wildness. "8. Fierce, savage, ferocious, furious, violent, destructive, cruel. 9. Of the sea, a stream, the weather, etc." Except for the solar system and galaxies, these entries pretty much suggest the whole world is wild, and that, in turn, should remind us of Thoreau's most famous saying: ". . . in Wildness is the preservation of the world."[2]

By the time he wrote these words, Thoreau had gone through a conversion from a wannabe literatus publishing regrettable verse and awkward translations of Greek tragedies to an inspired Humboldtian scientist. He had read Alexander von Humboldt's biography, and his books *Cosmos* (Humboldt, 1850) and *Personal Narrative of Travels to the Equinoctial Regions of America* (Humboldt, 1907), a scientifically rich account of his five-year sojourn in the Amazon and Central America.[3] He was aware of William Herschel's discoveries of deep space and time in astronomy, one of Ralph Waldo Emerson's obsessions, and the *Preliminary Discourse on the Study of Natural History*, the first treatise on scientific method, by William's son John Herschel (1845), and one of Emerson's favorite books (he could not stop recommending it to friends). He read the three-volume *Principles of Geology* by Charles Lyell (1833)—his first formal reading in natural history. As soon as *On the Origin of the Species* by Charles Darwin (1859) was published, Thoreau read it with approval.

The *Journals* by Thoreau (1861/2009) are the fruit of his conversion, a Romantic and transcendental tribute to the wonders of natural science. In Thoreau's famous statement, wildness is not an image or metaphor; it is a simple truth anchored in natural history and the physical sciences.

The final entries in the *OED* are primarily psychological, beginning with "10. Of feelings or their expression: Highly excited or agitated; passionately vehement or impetuous." Indeed, most of the last entries concern humans—*not having control of one's mental faculties*, or suggestive of the absence of control. They remind us that many, perhaps most, uses of the term wild refer to human beings (remember all those romance novel titles?), which in turn reminds us, forcefully, that human beings are wild. The prodigious extent of our artifice cannot mask that brute fact.

The last entry on the subject in the *OED*—it seems like an afterthought—reads: "wild (e, obs. ff. WEALD, WIELD *v.*; obs. pa. t. of WILL. *v*," and is surely the most arcane pronouncement ever made on our subject. I believe it's one of the most important, too, since it specifically connects the etymology of wild with the word will.

The *OED* definitions are reminders of the complexity of our concept, and a prod to tease out and lay bare this complexity. Wittgenstein (1953, p. 50e) says that the "work of the philosopher consists in assembling reminders for a particular purpose." That's a good start, but I want to do more.

Why should we be surprised at the relation between wild and will? Are we startled when in the second paragraph of *Wilderness and the American Mind*, Nash (1982), p. 1) writes: "In the early Teutonic and Norse languages, from which the English word in large part developed, the root seems to have been 'will' with a descriptive meaning of self-willed, willful, or uncontrollable," and "the Old English 'deor' (animal) was prefixed with wild to denote creatures not under the control of man."

This is also Thoreau's sense of the word.

Anyone familiar with Thoreau's writings knows that he makes broad claims for the word, such as in "all good things are wild and free" (Thoreau, 2007, p. 210) and "life consists with wildness" (ibid., p. 210). There are also passages where his meaning specifically mirrors Nash's paragraph: "Living wild—whatever has not come under the sway of man is wild—In this sense original & independent men are wild—not tamed &

broken by society"; or, "The most alive is the wildest. Not yet subdued to man." And most precisely, after reading *The Study of Words* by Richard C. Trench (1852), Thoreau (as cited in Paul, 1958, p. 412) comments in his *Fact Book*, "Wild—past participle of *to will*—self-willed."

These passages are as close as we will come to a satisfactory definition, but they are vulnerable to an obvious question: What does self-willed mean? Self is one of the most obscure words in our philosophical lexicon; will is perhaps even more problematic. Both are burdened with 2,500 years of contention and dispute. And self-willed? I think it's a philosophical abyss. Recall Milton (1668, lines 910–911): "Into this wild Abyss / The womb of Nature, and perhaps her grave." Let's not go there.

If we are going to preserve a useful sense of wildness as self-willed, we need to unpack the term self-willed and learn to talk about it in new ways. I will deploy three other rich concepts: autonomy, self-organization, and tzu-jan.

The Greek word *autos* means self. Combined with *nomos* (law), it yields the compound *autonomos*, which translates in English (beginning in the early nineteenth century) as *autonomous*. Like English, ancient Greek had many compound self-words—and modern science is littered with them: *automorphos*, self-formed; *autoretor*, self-made; *autobatheous*, self-supporting; and our friend, *autognotos*, self-willed. Recently Humberto R. Maturana and Francisco J. Varela (1992) coined the term *autopoesis*, for self-making, to designate the organization unique to living beings.[4] There are now volumes on the concept.

The common dictionary meanings of autonomous are "a country or region enjoying self-government," or, "an individual that acts independently and has the freedom to do so." It is also, of course, a key term in Immanuel Kant's moral philosophy—the basis of our dignity. These meanings refer mainly to social and philosophical issues concerning human beings (and perhaps some other animals; I certainly would include at least whales and dolphins). I will set them aside, not because they are unimportant, but because it would take us far from my main subject.

Autonomous refers also to a wide variety of natural systems that are clearly not subjects of human mastery or dominance. For instance, in *The Creation*, Wilson (2006, p. 92) reminds us that:

Even if Earth's threatened biodiversity in all it immensity could be reanimated and bred into populations awaiting return to what might in the twenty-second century pass for the "wild," the reconstruction of thereby independently viable populations is beyond reach. Biologists haven't the slightest idea of how to build a complex autonomous ecosystem from scratch.

Or consider Princeton biologist Simon Levin's essential features of complex adaptive systems, outlined in *Fragile Dominion: Complexity and the Commons*. The third of these characteristics is "*An autonomous process* (such as natural selection) that uses the outcomes of . . . local interactions to select a subset of those components for *replication* or *enhancement*" (Levin, 1999, p. 12). Put simply, evolution is an autonomous process. Levin also makes favorable mention of ecosystems and the biosphere, and is friendly toward the scientific (as opposed to the New Age) theory of Gaia proposed by James Lovelock (2000) and Lynn Margulis (1999).

Not all autonomous systems are external to our bodies. There is the autonomic nervous system, one of the oldest parts of our nervous system and one not subject to volitional control. Our hearts pump, our lungs breathe, and our synapses fire—and we don't have to do any of it. Perhaps most interesting is the autoimmune system, whose fundamental characteristic is that it distinguishes self from other; it is a system that must both identify each challenge to the self, interior or exterior, and respond, autonomously. Even when the immune system fails, the language of *autos* persists; when both the identification and response functions break down, the result is called *autoimmunity*, and the ensuing illness is labeled *autoimmune disease*.

A long and distinguished roster of researchers has attempted to explain our immune system, but it remains a mystery. Alfred I. Tauber (1997), professor of philosophy and medicine at Boston University, has written a critical history of immunology: *The Immune Self*. (He has also written a remarkably innovative book on Thoreau, *Henry David Thoreau and the Moral Agency of Knowing* [Tauber, 2001], and interestingly, a book on the autonomy of patients [Tauber, 2005]). Tauber has done as much as anyone to develop a workable definition of what biologist are now wont to call *the biological self*—but in the end its bounds remain as messy and uncertain as those of the psychological self. What's clear is that we do not control our immune system—or our organs, nervous system, or the firing of synapses. They do their work, and like all complex systems are adaptive.

Recent research is expanding the way we need to think about the self, my singularity—*me*, my *subjectivity*, as they say, the *thing* that experiences, my *individuality*—affirming that it's far more of a community effort than we've typically thought. We harbor multiple populations of bacteria, microbes, an entire microbiome, and *they* (I am no long inclined to say *it*), are, it seems, responsible for *me*, as it were. It's beginning to be hard to talk about these developments.

There are ten times as many bacteria in and on us—our teeth, the residual web between our toes, our gut, our eyelids, everywhere—than there are human cells in our body, and they are wildly autonomous, active ecosystems nested and integrated within the larger ecosystem we designate as *me*. Between 500 and 1,000 species of these bacteria have been discovered, and we know very little about the majority of them except that they are essential to our functioning and health—and that somehow they negotiate the balance between self and other.

I think of their presence as the most recent variation on the theme that human beings don't occupy a privileged niche—that we don't live at the center of the universe, that we aren't God's children or independent agents dignified by Kantian freedom, but rather are, as a recent article in the *New Yorker* (Bilger, 2010, p. 106) put it, "walking, talking microbial vats." In the same article, with no exaggeration, Margulis is quoted as saying, "There is no such thing as an individual. . . . What we see as animals are partly just integrated sets of bacteria" (ibid., p. 106). That includes us, obviously. As a community of autonomous microbes, each of us retains our biological wildness, despite our ideologies of control, antibiotics, radiation treatments, vaccines, and many other intrusions. We simply have not come to grips with how much we owe our microbes. Indeed, a recent summary of our personal bacteria in the *New York Times* was titled "How Microbes Defend and Define Us" (Zimmer, 2010). To paraphrase Thoreau, *in bacteria is the preservation of me.*

As I mentioned above, compound self-words are common in science—self-replicating, self-adapting, and so forth. Of these, the concept of self-organization is perhaps the most widespread, influencing disciplines as varied as astrophysics, microbiology, and the mathematics of form. Kant used it in his *Critique of Judgment* (1978) (see part II, chapter 4, passim) to distinguish between an organism, in which self-reproducing parts are the means to one another and exist for one another, and a machine, in which the parts are not self-reproducing and exist together only after assembly for a humanly established function. That remains a difference that makes a difference.

Twentieth-century physics left us with a formidable anomaly: for large-scale objects, relativity rules; for small-scale ones, quantum theory rules. The problem is that no one has yet figured out exactly how they are related. Two years ago, three physicists wrote an article for *Scientific American* suggesting a theory of quantum gravity that would show how space and time, the basic building blocks of the cosmos, "pull themselves together"—the self-organizing quantum (Jurkiewicz, Loll, & Ambjom,

2008). The Fermi Institute at the University of Chicago researches magnetic self-organization in interstellar plasma. Other scientists study self-organization in crystallization and thermodynamic systems. Chemists explore molecular self-assembly, while biologists look at the spontaneous folding of proteins, the self-organization of social structures in insect populations, self-organization in flocks of birds and schools of fish, and the origin of life. On this last subject, *The Origins of Order: Self-Organization and Selection in Evolution* by Stuart A. Kauffman (1993) remains the classic statement. And of course, self-organization and self-maintenance are central to the Lovelock-Margulis theory of Gaia.

I've been particularly struck by self-organization in the more ephemeral aspects of nature. *The Self-Made Tapestry: Pattern Formation in Nature* (Ball, 1999) describes patterns, visible to us daily—bubbles, waves, logarithmic spirals, branches, fluids, blooms, grains, and communities. (The variety of logarithmic spirals, involving self-similarity, is marvelous—galaxies, bighorn sheep horns, hurricanes, the chambered nautilus, pinecones, the heads of sunflowers, and the seahorse tip of the Mandelbrot set.) Consider these broad categories and then imagine, if you can, the multiple, particular, self-organized phenomena surrounding us—at all levels. This includes social organization; research projects in the self-organizing properties of human groups and economic behavior are legion—too numerous to mention here.

All this brings home the gross inadequacy of the reductive statement that "everything is connected to everything else." We cannot escape from a world of particulars organized around their self. Individuation and connection both strike deep into reality, but we have emphasized things and their connections at the expense of process, and hence our view of reality lacks balance. Like the ancient Chinese and the nude, we missed something. We've seen the duck; now we need to see the rabbit. Fortunately, the ancient Chinese did not miss the rabbit.

The term tzu-jan is taken to mean both *natural* and *spontaneous*. It is also a philosophical term of considerable importance in Taoism, appearing in the Lao-tzu's *Tao Te Ching* and the writings of Chaung-tzu, among many other texts. The translator David Hinton (2000, pp. xx–xxi) provides an introduction to its complexity:

Tzu-jan's literal meaning is "self-so" or "the of-itself," which as a philosophical concept becomes "being such of itself," hence "spontaneous" or "natural." But a more revealing translation of tzu-jan might be "occurrence appearing of itself," for it is meant to describe the ten thousand things [a common Taoist and Buddhist metaphor for everything; e.g., see *Tao Te Ching*, section 34] burgeoning forth spontaneously from the generative source, each according to its own nature, independent and self-sufficient, each dying and returning to the process of change, only to reappear in another self-generating form.

Others translations of tzu-jan are "what is spontaneously so" (Ames & Hull, 2003, p. 115), "that which is so by itself" (Hahn, 2001, p. 204), and "each new 'so' as it hatches 'on its own'" (Jullien, 2009, p. 164).

A great deal depends on how we think of tzu-jan. We glimpse a sense of its significance and complicated relation to the tao (usually translated as "the Way") in section 25 of the *Tao Te Ching*:

Human abides by earth
Earth abides by heaven
Heaven abides by Way[5]
Way abides by occurrence appearing of itself [tzu-jan].
(Hinton, 2000, p. 28)

There is no doubt of its primacy in Chinese thought about nature, whatever the intricacies of its translation. The classic essay "The Continuity of Being: Chinese Visions of Nature" by Tu Wei-ming (1989) opens with this quotation from the equally classic *Intellectual Foundations of China* by Frederick M. Mote (1971, pp. 17–18): ". . . they [the Chinese] have regarded the world and humans as uncreated, as constituting the central features of a spontaneously self-generating cosmos having no creator, god, ultimate cause, or will external to itself." No will external to itself—shall we say, then, self-willed? Here may be Thoreau's central meaning of wild, again.

Lest you think this excursion into the concept to tzu-jan is an indulgence in Taoist metaphysics, consider the way that two of the world's most distinguished physicists, Stephen Hawking and Leonard Mlodinow (2010, p. 180), end their recent survey of modern physics, *The Grand Design*. After the cosmological constant, bosons, Feynman paths, black holes, string theory, m-theory, Conway's game of life, and "the theory of everything," in the antepenultimate paragraph, we find this—"Spontaneous creation is the reason there is something rather than nothing, why the universe exists, why we exist. It is not necessary to invoke God." Nor for that matter is it necessary to invoke the creation, prime mover, life force, spirit and all their kin. Universes, of which there turn out to be many more than one, just happen.

In a deep sense, we have come full circle in our Western thinking about nature, if only we can realize and appreciate it. *Physis* is the Greek word for what we now call physics. Its root means "growth." Similarly, the

Latin root of *natura* means "birth." And both terms suggest an organic conception of the cosmos sympathetic to Taoist accounts. Martin Heidegger's scrutiny of pre-Socratic Greek philosophy includes an extensive examination of physis. In his *Introduction to Metaphysics*, Heidegger (2000, p. 15) asks:

Now what does the word *physis* say? It says what emerges from itself (for example, the emergence, the blossoming, of a rose), the unfolding that opens itself up, the coming-into-appearance [so similar to Hinton's translation of tzu-jan] in such unfolding, and holding itself and persisting in appearance—in short, the emerging-abiding sway.

It appears virtually impossible to talk about physis, natura, or wildness without including self-words.

Heidegger believed that nature in the narrow sense of (what became historically) the physical sciences derived from the experience of physis in poetry and thought. He notes that "*physis* as emergence can be experienced everywhere: for example the celestial processes (the rising of the sun), in the surging of the sea, in the growth of plants, in the coming forth of animals and human beings from the womb." Yet he goes on to say that physis is not just one process among many but rather "is Being itself, by virtue of which beings first become and remain observable" (ibid.). My various elaborations of wildness also suggest a ceaseless process of the ever new, the generative power of autonomous processes and self-organization. The wild is not static.

Although he did not express it in terms of tzu-jan or physis, Thoreau was certainly aware of this. In *Walden*, he observes: "There is an incessant influx of novelty into the world, and yet we tolerate incredible dullness" (Thoreau, 1971, p. 332). Within that influx of novelty there is order—the order recorded by the natural sciences; the order that is part of the Greek meaning of the word cosmos. (In "Walking," Thoreau says that "the Greeks called the world κόσμος, Beauty—or Order, but we do not see clearly why they did so, and we esteem it at best only a curious philological fact." [Thoreau, 2007, p. 217]) This incessant flux or process is not separate from its spontaneous manifestations, and its manifestations are not separate from it, assuming we can use the word manifestation without thereby implying that there is something behind it that manifests. There is no dualism here—something that Heidegger and the Taoists are striving to say, but that perhaps cannot be said. (*Wovon man nicht sprechen kann, darüber muss man schweigen.*) Being (whatever Heidegger meant by that), physis, tzu-jan, Hawking and Mlodinow's spontaneous creation, the self-willed, and the wild are permanently in the ceaseless world of impermanence, and in our experience of them; there is no separating out that spontaneity or wildness from its coming forth.

In short, the objects and events that we think of as constituting our world, are the instantaneous products of an active processional force, but there is no reified Platonic form out there for us to find; all we will ever have is its display.[6] Without this force, the cosmos would not renew itself. Everything—galaxies to bacteria—surges forth, continuously arising and passing away, productive and fluid, without a preordained expression of order, and yet, again, it is not without *some* order. (The mysterious relation between flux and order challenged both the Greeks and Taoists.) Each spontaneous display is original: the world has never been this way before, or the mind either. Each may very well be caused, and the whole process deterministic, yet how could we know the causal roots unfathomably buried in the past 13 billion years? What I know is that each moment of my experience I am left with the poignancy of memory and a cosmos forever changed.

Are then wildness, self-will, tzu-jan, and spontaneous creation the same thing? No. They are each particular cultural expressions, although they have an affinity for each other and perhaps even, in Wittgenstein's sense, a family resemblance. And each can, I believe, enable a broader understanding of the wild. Each opens to thought and experience the stunning pervasiveness of wild nature in our lives. We are embedded in it, permeated by it, saturated with it, constructed by it, and maintained by it; we owe our existence to it. The idea that we have to go to Yellowstone to find or experience the wild is absurd. We simply must learn to switch scales with ease, moving deftly from the vastly big to the infinitesimally small, integrated hierarchies. The memorable film *Powers of Ten*, directed by Charles Eames and Ray Eames (1998), helps. So do the various newer Web versions of the film, and the various books based on it. Wildness is not limited to the human scale, our love for large fauna, and what we can see. We need to get over that. Study everything from quarks to worms to the clusters of galaxies portrayed in the Hubble Deep Field images; everything we can find, everywhere, is coming forth, spontaneously, moment by moment.

I think of this pervasive wildness as the Vast Array. There are approximately 150 billion galaxies, trillions and trillions of stars, and probably trillions of planets. These galaxies are expanding at near the speed of light. As Terrence Dickinson (2006, p. 20) puts it, "In the time it takes to read this sentence, the universe will increase in volume by 100 trillion cubic light years." Whee! Gazillions of electromagnetic waves zing about. How many? A single one-watt night-light emits a billion billion (not a typo) photons per second (Hawking and Mlodinow 2010, p. 70). At each moment of a person's experience, billions of neurons and a trillion synapses are waiting, analyzing, and firing—even while we sleep. Air contains an unimaginable stew of beings—viruses, bacteria, seeds, spores, pollen grains of fungi, algae, mosses, and liverworts as well as caterpillars, aphids, spiders, butterflies, moths, beetles, and mites. A single puffball produces 7 trillion spores (Lukas, 2003). Bacteria are ubiquitous and can travel thousands of miles. Walk up a mountain and you inhale a cornucopia of all this stuff. Then the bacteria in your lungs—2,000 of them per square centimeter—sort, absorb, and signal the bacteria in your immune system to launch a SWAT team and kill any bad guys (Zimmer, 2010). And you don't even know. A quart of seawater off Maui contains 10 billion viral particles. Marine viruses kill 100 million metric tons of microbes *every minute* (Judson, 2009). The debris feeds the bottom of the food chain. Microbes' DNA may be the most prevalent on Earth. The Oxford entomologist J. M. Anderson says that temperate forest soil can "support up to a thousand species of soil animals . . . present in populations exceeding one to two millions per square meter" (Louma, 1999). And the endless swarms of predator choanoflagellates? Don't forget them. Everywhere we look, we encounter multitudes of fellow wild beings. You want Thoreau's "*Contact*! *Contact*! *Who* are we? *where* are we?" (Thoreau, 1972, p. 71). Contact with the wild? You've got it, right here and now. There is no escape.

Why in the world must something as compelling and readily observable as this require elaboration? Why hasn't the greater part of humanity noticed it, attended to it, and *absorbed* it? Why aren't we collectively stunned with appreciation?

Part of the answer is that neither the general public nor even most of the environmental community appreciates the wild, self-willed, autonomous, self-organizing aspect of the physical sciences. Some scientists do, of course, but the environmental community is obsessed with policy, legislation, and activism; it's busy looking at screens, talking on phones, and funding all of it—in a word, politics. Too many of us have become geeks, wonks, and bloggers who spend precious little time out and about in wild nature, or contemplating its meaning. Thoreau's tonic is ignored. How many people in the employ of environmental organization are expert in the particulars of any natural science? How many of them know, really know, bird eggs, double stars, or alpine plants? What do they know? Politics.

Then too, as a people, we are largely ignorant of science and mathematics. Witness our scores on international exams: the United States ranks twenty-third in science and thirty-first in mathematics—well behind Estonia, Slovenia, and Slovakia. Or consider the data on evolution. According to a June 2007 *USA Today*/Gallup poll, only 18 percent of Americans believe that evolution is "definitely true." A Pew Forum poll offering the proposition "life on Earth has evolved over time by evolution through natural selection" found a mere 26 percent of Americans in agreement. In a European poll asking if "human beings, as we know them today, developed from earlier species of animals," the proportion of yes answers varied from 85 percent agreement in Iceland down through thirty countries to 27 percent agreement in Turkey, so it looks like we might rank below Turkey in this respect (Dawkins, 1999, p. 433). Roughly a third of Americans know about the big bang theory—a figure so low that the embarrassed National Science Foundation removed the question from its annual survey of Americans' knowledge of science (Phelps, 2010). This is discouraging.

As a people, we have little intellectual or emotional engagement with science or mathematics. Despite numerous television specials, videos, and the coffee table books collecting everything from Hubble photographs to portraits of our synapses, to an astonishing degree the rudiments of physical science have yet to sink into the public consciousness. As the Nobel Prize physicist Richard Feynman (1955) noted, "This is not yet a scientific age."

The challenge for appreciating tzu-jan is slightly different. Our culture has inherited a nature of objects and events existing in Euclidean space with time as an independent variable. Few of us are well attuned to the process of a self-willed world, and hence the need for elaborations. Even a slight acquaintance with the multifaceted world of other cultures reveals that we are working against conceptual barriers to appreciating wildness as a process. As our own ways of thinking threaten to become "locked in," we need to open up to different ways of thinking. The experience of tzu-jan is free, though difficult; it is a matter of attention and practice.

Increasingly, scholars are probing the line between Western and Chinese thought. For instance, Jullien (2004, p. 48), trained in Western classics and a first-rate sinologist, says, "The Chinese perspective opens directly onto the coherence inherent in change itself and so bestows on becoming the logic of its own transpiring." That is not easy to understand.

We have, perhaps *it* has, an ability "to maintain itself ever in process" (ibid.), but we are not there yet; many of us are unaware of its existence. Walking that path, we are available, as it were, to an *experience* of raw, wild, ceaseless process—but only available. We cannot will to remain there. The wild may surge forth spontaneously, *being* the cosmos and the mind, but it seems we cannot, at present, remain in its company. *This is not yet a wild age.*

Conclusion

I imagine that in the past, entire generations lived "ever in process." Being again in the presence of it may help us inhabit that impermanent place again, emotionally and intellectually. Some of the authors in *American Earth* suggest this is so: Thoreau, of course, Henry Beston in the dunes, Rachael Carson beside her tide pool, Wilson with his ants, Annie Dillard at her creek, Terry Tempest Williams on the slickrock, and Mary Oliver with her grasshopper. However abstract the ultimate analysis of wild nature, we can only experience its spontaneous presence. It is no accident that they—and Abbey, Snyder, and Nash—were intimate with natural history, and spent some of their lives in wild places, often in relative solitude. That potent brew of awe and wonder inaugurated our literature of the wild. Now we must elaborate it. So we walk on, face to a hard wind.

Notes

1. George Dyson, "A Universe of Self-Replicating Code." (www.edge.org/conversation.php-a-universe-of-self-replicating-code); Drew Endy, "Engineering Biology" (www.edge.org/conversation.php?cid-engineering-biology); Dennis Overbye, "Mystery of Big Data's Parallel Universe Brings Fear, and a Thrill," *New York Times*, 4 June 2012.

2. See Thoreau (2007), p. 202; *Atlantic*, June 1862.

3. For Thoreau's reading, see Sattelmeyer (1988). Note that Robert Sattelmeyer lists only books for which there are records—libraries, including his personal library, journal records, correspondence, and so on. No doubt Thoreau read many other books for which there is no formal record. For instance, he had the run of Ralph Waldo Emerson's vast library. For Thoreau, Humboldtian science, and the Herschels, see Walls (1995, 2003, 2009). Laura Dassow Walls's work has transformed studies of Thoreau and American transcendentalism. In a similar vein, *The Age of Wonder: How the Romantic Generation Discovered the Beauty and Terror of Science* by Richard Holmes (2008) confirms British Romanticism's infatuation with science. Lord Byron, Percy Bysshe Shelley, Samuel Taylor Coleridge, Emerson, and Thoreau welcomed science, and found it no threat to the humanities.

4. The concept was originally formulated in various essays.

5. In *Tao Te Ching: A Book about the Way and the Power of the Way*, Ursula K. Le Guin (1997, p. 35) translates tzu-jan in section 25 as "what is." In her commentary on this section, she notes, "I was tempted to say, 'The Way follows itself.' Because the Way is the way things are; but that would reduce the significance of the words. They remind us not to see the Way as a sovereignty or a domination, all creative, all yang. The Way itself is a follower. Though it is before everything, it follows what is" (ibid., pp. 34–35.). I do not find that particularly coherent. "What is" lacks the active force described by Hinton. It seems static, and tzu-jan is not static. In *Taoteching* by Red Pine (1996, p. 50), the last line of section 25 is translated as "The Tao imitates itself." I really dislike the verb imitates, because of its implied dualism, and because of its association with simulation and copy. I see tzu-jan as being original.

6. Processional is a term used by Roger Ames and David Hall in their various translations of Asian classics.

References

Ames, R., & Hull, D. L. (2003). *Dao De Jing: A philosophical translation*. New York, NY: Ballantine Books.

Ball, P. (1999). *The self-made tapestry: pattern formation in nature*. New York, NY: Oxford University Press.

Bilger, B. (2010). Nature's spoils. *The New Yorker*, November 22, 2010.

Clark, K. (1972). *The nude: A study in ideal form*. Princeton, NJ: Princeton University Press.

Commoner, B. (1971). *The closing circle: Nature, man, and technology*. New York, NY: Random House.

Darwin, C. (1859). *On the origin of the species by means of natural selection, or the preservation of favoured races in the struggle for life*. London, UK: John Murray.

Dawkins, R. (1999). *The greatest show on earth: The evidence for evolution*. New York, NY: Free Press.

Dickinson, E. (1961). In T. H. Thompson (Ed.), *Final harvest: Emily Dickinson's poems*. Boston, MA: Little, Brown. (Original work published 1890).

Dickinson, T. (2006). *NightWatch: A practical guide to viewing the universe*. Buffalo, NY: Firefly Books.

Dyson, G. (2012). A Universe of Self-Replicating Code. Retrieved from http://www.edge.org/conversation.php-a-universe-of-self-replicating-code.

Eames, C., & Eames, R. (Directors). (1998). *Powers of ten: A flipbook* New York, NY: W. H. Freeman.

Endy, D. (2008). Engineering biology. Retrieved from http://www.edge.org/conversion.php?cid-engineering-biology.

Feynman, R. (1955). The value of science. *National Academy of Sciences* (Fall). Retrieved from http://www.phys.washington.edu/users/vladi/phys216/Feynman.html.

Hahn, T. H. (2001). An introductory study on Daoist notions of wilderness. In N. J. Girardot, J. Miller, & L. Xiaogan (Eds.), *Daoism and ecology* (pp. 201–218). Cambridge, MA: Harvard University Press.

Hawking, S. & Mlodinow, L. (2010). *The grand design.* New York, NY: Bantam.

Heidegger, M. (2000). *Introduction to metaphysics.* New Haven, CT: Yale University Press.

Herschel, J.F.W. (1845). *Preliminary discourse of the study of natural history.* London, UK: Brown, Green, Longmans.

Hinton, D. (2000). *Tao te ching.* Berkley, CA: Counterpoint.

Holmes, R. (2008). *The age of wonder: How the romantic generation discovered the beauty and terror of science.* New York, NY: Pantheon.

Judson, O. (2009). A gazillion tiny avatars. *New York Times,* December 15. Retrieved from http://opinionator.blogs.nytimes.com/2009/12/15/a-gazillion-tiny-avatars.

Jullien, F. (2004). *In praise of blandness: Proceeding from Chinese thought and aesthetics.* New York, NY: Zone Books.

Jullien, F. (2007). *The impossible nude: Chinese art and Western aesthetics.* Chicago, IL: University of Chicago Press.

Jullien, F. (2009). *The great image has no form, or on the nonobject through painting.* Chicago, IL: University of Chicago Press.

Jurkiewicz, J., Loll, R., & Ambjom, J. (2008). Using causality to solve the puzzle of quantum spacetime. *Scientific American, 299* (July) 1, 42–49.

Kant, I. (1978). *Critique of judgment.* New York, NY: Oxford University Press.

Kauffman, S. A. (1993). *The origins of order: Self-organization and selection in evolution.* New York, NY: Oxford University Press.

Le Guin, U. K. (1997). *Tao te ching: A book about the way and the power of the way.* Boulder, CO: Shambhala.

Levin, S. (1999). *Fragile dominion: Complexity and the commons.* Reading, MA: Perseus Books.

Loeffler, J. (1989). *Headed upstream: Interviews with iconoclasts.* Santa Fe, NM: Sunstone Press.

Loeffler, J. (2003). *Adventures with Ed: A portrait of Abbey.* Albuquerque, NM: University of Mexico Press.

Louma, J. R. (1999). *The hidden forest.* New York, NY: Henry Holt.

Lovelock, J. (2000). *Gaia: A new look at life on earth.* New York, NY: Oxford University Press.

Lukas, D. (2003). Of aerial plankton and the Aeolian zones. In D. Rothenberg & W. J. Pryor (Eds.), *Writing on air* (pp. 39–43). Cambridge, MA: MIT Press.

Lyell, C. (1833). *Principles of geology* (1st ed.). London, UK: John Murray.

Margulis, L. (1999). *Symbiotic planet: A new look at evolution.* New York, NY: Basic Books.

Margulis, L., & Sagan, D. (1995). *What is life?* Berkeley, CA: University of California Press.

Maturana, H. R., & Varela, F. J. (1992). *The tree of knowledge.* Boulder, CO: Shambhala.

McCarthy, C. (1985). *Blood meridian.* New York, NY: Random House.

McKibben, B. (Ed.). (2008). *American earth: Environmental writing since Thoreau.* New York, NY: Literary Classics of the United States.

Milton, J. (1668). *Paradise lost.* London, UK: Creed Church.

Mote, F. W. (1971). *Intellectual foundations of China.* New York, NY: Alfred A. Knopf.

Nash, R. (1982). *Wilderness and the American mind* (3rd ed.). New Haven, CT: Yale University Press.

Overbye, D. (2012). Mystery of big data's parallel universe brings fear, and a thrill. *New York Times,* June 4.

Paul, S. (1958). *The shores of America: Thoreau's inward exploration.* Urbana, IL: University of Illinois Press.

Phelps, T. (2010). National Science Foundation removes America's ignorance of evolution and the big bang from report. *Examiner,* April 9. Retrieved from http://www.examiner.com/article/national-science-foundation-removes-america-s-ignorance-of-evolution-and-the-big-bang-from-report.

Pine, R. (1996). *Taoteching.* San Francisco, CA: Mercury House.

Sattelmeyer, R. (1988). *Thoreau's reading: A study in intellectual history with bibliographical catalogue.* Princeton, NJ: Princeton University Press.

Snyder, G. (1990). *The practice of the wild.* San Francisco, CA: North Point Press.

Tauber, A. J. (1997). *The immune self: Theory or metaphor?* New York, NY: Cambridge University Press.

Tauber, A. J. (2001). *Henry David Thoreau and the moral agency of knowing.* Berkeley, CA: University of California Press.

Tauber, A. J. (2005). *Patient autonomy and the ethics of responsibility.* Cambridge, MA: MIT Press.

Thoreau, H. D. (1972). *The writings of Henry D. Thoreau: The Maine Woods.* J. J. Moldenhauer (Ed). Princeton, NJ: Princeton University Press.

Thoreau, H. D. (1971). *The writings of Henry D. Thoreau: Walden.* J. L. Shanley, (Ed.). Princeton, NJ: Princeton University Press. (Original work published 1854).

Thoreau, H. D. (1973). *The writings of Henry D. Thoreau: Reform papers.* W. Glick, (Ed.). Princeton, NJ: Princeton University Press. (Original work published 1849).

Thoreau, H. D. (2007). *The Writings of Henry D. Thoreau: Excursions.* J. J. Moldenhauer, (Ed.). Princeton, NJ: Princeton University Press.

Thoreau, H. D. (2009). *The journal, 1837–1861.* D. Searls, (Ed.). New York, NY: NYRB Classics. (Original work published 1861).

Trench, R. C. (1852). *On the study of words.* New York, NY: Redfield.

von Humboldt, A. (1850). *Cosmos: A sketch of a physical description of the universe* (Vol. 1). New York, NY: Harper and Brothers.

von Humboldt, A. (1907). *Personal narrative of travels to the equinoctial regions of America.* London, UK: George Bell and Sons.

Walls, L. D. (1995). *Seeing new worlds: Henry David Thoreau and nineteenth-century science.* Madison, WI: University of Wisconsin Press.

Walls, L. D. (2003). *Emerson's life in science: The culture of truth.* Ithaca, NY: Cornell University Press.

Walls, L. D. (2009). *The passage to Cosmos: Alexander von Humboldt and the shaping of America.* Chicago, IL: University of Chicago Press.

Wei-ming, T. (1989). The continuity of being: Chinese visions of nature. In J. B. Callicott & R. T. Ames (Eds.), *Nature in Asian traditions of thought* (pp. 67–78). Albany, NY: State University of New York Press.

Wilson, E. O. (2006). *The creation: An appeal to save life on earth.* New York, NY: W. W. Norton.

Wittgenstein, L. (1953). *Philosophical investigations.* Oxford, UK: Blackwell.

Zimmer, C. (2010). How microbes defend and define us. *New York Times,* July 12. Retrieved from http://www.nytimes.com/2010/07/13/science/13micro.html.

3

The Old Rules

Elizabeth Marshall Thomas

We consider ourselves at the top of the evolutionary ladder. Of course we do. We invented the ladder. Who else would we put at the top? Never mind that every other life-form has abilities that either we lack entirely or share in a reduced manner. We walk on our hind legs, we talk, and we use tools. These are the features that matter. Other life-forms don't do these things, or not as well, and thus are not as advanced by our standards. This has not been a beneficial concept for our planet, and may yet bring the planet to its knees.

Yet for all the destruction our species is causing, the natural world is still functional and in some places flourishing. This is because, with the exception of human beings along with their domestic plants and animals, all living things continue in the Old Way, provisioned by the land or sea in one way or another, keeping the Old Rules as set by their species. Without these rules, life on Earth would never have started, let alone have continued for three billion years. Old Rule One demands that you do your best to stay alive. Old Rule Two demands that as custodian of your species' hard-won genes, you do your best to keep them going.

The rules might seem simple, but the ways in which they're kept can be complex. Consider Cordyceps, for example. These fungi are in no way unusual except for the way they handle Old Rule Two. A Cordyceps will lie on the ground, waiting for an insect to crawl over it. The fungus will attach itself to the insect and release a chemical that burns a hole in the insect's carapace. The fungus goes inside. But the hole makes the insect vulnerable to attack and infection, and the fungus needs the insect to perform a service. So the fungus produces an insecticide to fend off potential attackers, a fungicide to fend off other infecting fungi, and an antibiotic to fend off germs that might make the host sick. Meanwhile, the fungus has been consuming the host's nonvital organs. No doubt this helps the fungus to mature. When the host's nonvital organs are gone, the fungus enters the host's brain and causes the host to climb a tree. The host climbs until it can climb no higher. But now, the fungus is where it wants to be, so it kills the host, extrudes its fruiting body, and releases spores by the thousands that float off through the air.

All this, just so the young fungi can begin life in new places, far from the starting points of their parents. If Cordyceps could not accomplish these seemingly miraculous maneuvers, its young could not disperse, which might result in a competitive mass of closely related Cordyceps with inadequate nourishment and little chance of continued survival. In contrast, grass, while not a fungus, just makes tiny, rather dry seeds that the wind blows around without a problem. But if a Cordyceps tried this, it would evidently violate Old Rule Two.

As for Old Rule One—do what you must to survive—consider the tardigrades, the water bears, also called moss piglets (Margulis & Schwartz, 1988). Most kinds of tardigrades are so small as to be virtually microscopic. The ones I've seen were transparent, so that with a microscope I could watch their food as it moved around in their insides. But because they have two eyes and eight legs, they are believed to be relatives of mites, and thus do not belong to the kingdom Protoctista, as do most other transparent, microscopic, or almost-microscopic organisms, yet they are with us in the animal kingdom, although fully as humble there as Cordyceps is in the fungi kingdom. If a tardigrade were climbing up the point of a sewing needle, it would be thinner than the needle, and therefore could be hidden from view until the viewer turned the needle a bit and noted something miniscule stuck to it. Only a young person with good eyes could see it, though. An older person would need a magnifying glass.

Tardigrades don't seem to bother much with Old Rule Two. Unlike the spectacular reproductive arrangements of Cordyceps, a female tardigrade discards her eggs on underwater debris and moves on, or merely sheds eggs along with her outgrown cuticle. The tardigrades don't even need males, as their eggs need not be fertilized. If not, the eggs hatch as females who, bypassing a larval state, emerge in adult form, ready to give life to more females. If a male is involved, some of the offspring will also be males, but considering the arrangement, it's easy to see why most tardigrades are females.

How can tardigrades be so casual about Old Rule Two? Perhaps because they have so fully, so amazingly, complied with Old Rule One. Their various species survive worldwide in all kinds of places—some live in hot springs, and others in the Arctic—and although creatures so tiny don't often leave fossils, it is believed that their phylum has been around since the middle Cambrian period, or in other words, for about 200 million years.

How could such tiny, seemingly fragile creatures manage this? Well, they can survive at extreme temperatures—from 304°F to minus 454°F. Although 304°F is far above the boiling point of the water in which many tardigrades live, minus 454°F is just three degrees above absolute zero, the point at which all thermal energy is gone. Obviously, the climate variations of our planet do not trouble the tardigrades.

Actually, their observance of Old Rule One might seem excessive. For some reason, tardigrades can survive massive doses of radiation. It's hard to imagine how or why they acquired this ability, but they can survive up to 570,000 roentgens, or in other words, a thousand times more than would be lethal to a person. Astronauts took note of this, and in 2007 brought tardigrades into space, where although the tardigrades were not protected, most survived without a problem. Some appeared to succumb to intense ultraviolet radiation, but recovered when they returned to Earth and were placed in a damp environment. How did they do this? If a tardigrade is damaged or starved, it can form itself into a cyst, a tiny roundish lump, in which it contracts its little body and repairs itself. The radiated tardigrades went on to lay eggs that hatched.

But these cysts are just one method of tardigrade survival. Most tardigrades live in swamps or damp mosses, and all of them eat liquids—some by sucking juice from plants, and others by sucking juice from other tiny aquatic creatures including other tardigrades—and hence they have a certain dependence on liquid. To deal with this in case the liquid disappears, they withstand even the most serious droughts by turning into forms known as "tuns" because they look like tiny wine barrels. These tuns, like dust particles, ride the winds to faraway places, which, with any luck, will be better than the places they left. In the form of a tun, a tardigrade can live for about 100 years.

When one considers the abilities of the Cordyceps and tardigrade, our much-touted abilities of language, tool use, and hind-leg walking seem modest. In light of the fact that a single bacterium can perform feats far beyond our abilities, such as disintegrating a bit of oil that we spill in the ocean, and that bacteria have been on this planet for more than 3 billion years, we should realize that the ladder we think we're on top of is a human illusion. We share our planet with an estimated 20 or 30 million known life-forms, or roughly 1.7 million not counting bacteria. There are 62,000 different kinds of vertebrates, 1.3 million different kinds of known invertebrates (and more that are as yet unknown), and 300,000 different kinds of plants, not to mention roughly 60,000 different kinds of other life-forms such as fungi, lichens, and algae, which of course are important to life on Earth in many different ways. And every one of these 1.7 million life-forms has its own evolutionary ladder, and remarkable abilities.

Thus, it seems strange that most of us are unaware that we share the planet with so many different kinds of organisms, and tend to acknowledge only the big ones, and at that, rather vaguely—we know if it's a bear, frog, bug, tree, or bird. We're fairly sophisticated if we can identify them by their common names—European brown bear, Blanchard's cricket frog, green darner dragonfly, red oak, or white-throated sparrow—but with the exception of our domestic plants and animals, we have little concept of who they are or how they manage. As for the most significant ones, the tiny ones that operate the planet—helping plants grow as well as producing oxygen along with just about everything else except sunlight and water that keeps the rest of us alive—we don't even know they exist.

I became aware of the large gap between the realities of the natural world and our limited human concept of them due to an event involving a leopard in Uganda and a female psychoanalyst in Manhattan. Thankfully, I was not the patient of this psychoanalyst, nor would I have wanted to be. Despite the risks, I felt more affinity with the leopard.

What happened was this. I was doing fieldwork in northern Uganda, making a little study of the pastoral Dodoth people (Thomas, 1965). My son and daughter, ages two and three, respectively, were with me. We were camped near the settlements of several Dodoth families, with a vast, uninhabited bushland all around. One evening I accompanied some of the Dodoth to a dance in a settlement two or three miles away, but not until after dark did I learn that the people who came with me were planning to spend the night. This meant I would be going back to camp by myself. Since I had been counting on my companions to find the way home, I hadn't brought a flashlight or, of course, a weapon, but I thought, oh well, and started walking.

The moon was low in the west, the night was dark, and I had miles to travel. I got lost in the heavy bush, but with the help of the constellation Orion I was trying to go east, hoping to cross a north–south track made earlier by my vehicle, when instead I crossed a footpath. I stopped and looked around, and noticed the outline of a hill against the sky. I thought I recognized it. My camp might be on it, the path led toward it, and without a better idea of where to go, I started up the path. If it wasn't the right hill, I could always keep going east in search of the vehicle track.

Then, just ahead of me, a leopard coughed. This was worrisome. Evidently I was walking toward him. But since I didn't know where I was except that I was near a leopard, I had no idea where else to go. I didn't want to stand still in the dark, in case, not hearing me, he might wonder what I was doing and come nearer to find out. I didn't want to vocalize because I had no idea what he would make of it, and also because a woman's voice is not effective in a situation of that kind. When speaking to big cats in the wild, a loud, strong, male voice is substantially better—less like a helpless whine, and more like a dangerous roar. I once was able to soothe a bear by speaking to him gently, but he was scared, and I reassured him, which was not a message I wanted to send to the leopard. I also didn't want to leave the path or go anywhere with a leopard behind me, because big cats attack from behind. At least he was trying to tell me where he was. Hoping that I was right about the hill and my camp, I took the only option left to me and kept walking slowly toward him.

He coughed again, louder. I took this to mean: *What are you doing? I'm a leopard.* Well, I certainly knew he was a leopard, but could think of no other plan except to keep going. I believe he became exasperated, because after a moment of what seemed like astonished silence as he listened in disbelief to my slowly advancing footsteps, he coughed three times, as fast and forcefully as he could. *You're not listening! Now you're too close!* But I still didn't know what else to do except keep going.

He didn't cough again, so perhaps he thought that his efforts were useless and gave up. As I approached the place where he'd been coughing, I got the feeling that he had left. If he hadn't, at least he lay low and let me pass safely.

Why does this matter? Well, it mattered to me, certainly. He never would have coughed if he had been hunting me, in which case I'd be nothing but a dried-up leopard scat by now. Surely he coughed so that I wouldn't come up on him, get scared, and take the offensive in close quarters. For all their predatory skills, leopards are cautious. He surely wanted some distance between himself and a potential passerby.

Thus he was retaining an important part of Old Rule One as it applies to cats (Thomas, 1994). Do not fight and risk injury for no reason, as an injury will complicate your efforts for survival. You can warn and threaten, but don't attack unless you must.

Animals have all kinds of methods for avoiding unnecessary confrontations. His method, it seemed, was to alert me to his presence. When I didn't seem to understand, he took the initiative and left.

My camp was on that hill, and I reached it in a fairly short time. That's because the camp was near, not because I ran, as there is nothing worse than running from a predator unless you're certain you can get to safety before the predator gets to you. Better to show confidence, as if you felt so strong and competent that the predator is of no concern. If you can't do that, then it's often best to face the predator and try to make yourself seem bigger than you are. This is why our skin prickles when we're scared. We're raising our hair, although we don't know this consciously. Deep down we know it, though, and do as we would have done when our body hair was long and thick, back when we lived in the trees. All this is a crucial part of Old Rule One for primates and many other creatures, and I find it quite touching that we humans, with our tiny, inconsequential body hairs, continue to do it. We can hardly see the hairs ourselves, yet somehow we still believe that if they stand on end, they will make us seem big enough to threaten a predator.

As for the leopard, I think we already knew him. Most likely he was the same individual against whose nocturnal visits we had built a high thorn fence. Frequently at night I'd shine a flashlight around just to see what was out there, and sometimes (wow!) I'd see a pair of large, green, wide-apart eyes.

Three Ugandan men were with me as helpers, as were three people from the United States—two family members and a photographer—and we all made sure when we went to bed that the fence was intact and the gate was shut, and that everyone could see and hear the other people, or at least see their tents and hear them if there was a problem.

When I came up behind this leopard, then, he was probably on his way to contemplate our camp. Except for a few chickens that we locked up at night, there were no domestic animals in or near our camp—only people. In that case, what drew him? If he wanted to prey on just anybody, I would, as has been said, have become a forgotten leopard scat, unable to produce this chapter. Surely it wasn't me that lured him. Could it have been my children?

This was a disturbing question, and brings me to the female psychoanalyst from Manhattan. We met her shortly after we finished our work and returned to the United States. Back in those days, US children were supposed to sleep separately from their parents or incur serious psychological damage, and when she learned that my children slept right beside me in the same tent, she was horrified.

I was horrified too. Didn't she know what a leopard was? Should I have left my tempting little children alone and undefended if a leopard was considering them? I kept Old Rule Two as carefully as I possibly could. If my children had lived the same way in the Ugandan bush as they would have done had we been at home, they, too, might be the dust of leopard scats by now, not the splendid adults who are the joy of my life.

At one time, say twenty-thousand years ago when even our species lived in the Old Way (Thomas, 2006) and kept the Old Rules, every human being on Earth would have understood about predators, and would have been as shocked as I was by the psychoanalyst. With her city ways and blindness to the realities of camping in certain parts of Africa, she was as far from her origins as it is possible to be. Now and then, perhaps between the moment when the doorman opened the door of her building for her, and the moment when she stepped into a taxi, the autumn sun would shine on her, transmitting an unnoticed message to her pineal gland that the daylight wasn't as strong as it had been all summer—a sign that winter was at hand. As a result, she might get depressed, or eat or sleep more than she'd been doing, which are all possible reactions to pineal information. Hence Mother Nature might influence her slightly—but very slightly. She became my paradigm for oblivion to the natural world.

The leopard was just the opposite. He knew everything he needed to know about the Old Way. The Old Rules were important to him. Because the livestock of the Dodoth pastoralists had a negative effect on the local herbivores, and because the Dodoth gardeners made war on the baboons who robbed their crops, his food supply was surely diminished (Thomas, 1965). Old Rule One would demand that he find alternative prey. Of course he wanted to learn about our camp and its occupants, so he watched us. And when I surprised him in the dead of night, his life might have been chancy enough without his tussling with a stranger and possibly getting hurt, so he tried to warn me away. It would seem he did everything right, and he became my paradigm for awareness of the natural world.

Am I too focused on the psychoanalyst? There are billions of other examples—more than billions—of such blind indifference to the realities of nature. I can think of two in particular. The first is a book group to which I belonged until the reading requirement was a novel in which a man rides a horse for days across a waterless desert. The man couldn't do this any more than he could drive a car across the country if all the gas stations were closed. The car would need gas, and the horse would need a drink. I pointed this out to the book group members. They assumed I felt sorry for the horse. Not so. The novel was not a fantasy, or not intentionally; it was just carelessly written. The other reading group participants didn't see this. Didn't they consider the natural world to be worthy of at least the most obvious realities? They didn't, and neither had the author. Discouraged, I left the group.

I am also reminded of a television program, widely recommended, about Steven Hawking's announcement that there must be life on other planets. The program needed visuals, and its producers evidently hired artists to portray what the life-forms might be. The artists' efforts were appalling. With 1.7 million earthly forms to consider, they imagined only trees and large mammals—probably the only ones that came to mind. One of these life-forms was a hippolike creature that walked on its cylindrical head. That couldn't happen either. The thing had four legs. It just bumped its cylinder down between steps as if it needed five legs, which it wouldn't. The artist, under pressure to produce something fantastic, happened to chance on this unfortunate image and went with it. But it wasn't convincing, just depressing, because it meant that the artist as well as everyone else involved in that production knew so little about their own planet.

In the 1950s, when I was in my late teens and early twenties, I was lucky enough to experience the Old Way firsthand, not via the tardigrades I used to watch under a microscope, and not even via a leopard who, thanks to his discretion, didn't kill me, but instead via my own kind, the Ju/wa Bushmen of Nyae Nyae in the Kalahari Desert, in what is now Namibia (Thomas, 2006).

I went with my parents (my mother was an anthropologist) into a wilderness that extended through northern Namibia, western Botswana, and southern Angola. About 120,000 square miles of this vast bushland were essentially unexplored. (My dad, among other things, was an explorer.) The experience was time travel, because there we met Ju/wa Bushmen, now sometimes known as the First People, whose ancestors, according to DNA studies, were the ancestors of us all, whose language evidently gave rise to all other languages, and who were living on the shores of seasonal lakes in encampments that archaeologists were later to find had been occupied continuously for 35,000 years or longer. The dig was discontinued at that point, but almost certainly the occupancy was longer. A key finding of the archaeologists was that over all that time, the material culture of these people was essentially unchanged. Old Rule One says that change is risky. Survival is difficult enough, so that methods that are tried and true are repeated through the generations. The Ju/wasi lived entirely by hunting and gathering, as our ancestors had lived for 150,000 years, without agriculture or domestic animals, and without fabric or manufactured items. Everything they had came from the veld.

I have written of our experience extensively (Thomas, 2006) and won't repeat myself here, but mention it because the Ju/wasi give a good picture of who we all were when our species began, living as hunter-gatherers of the African savanna in the Old Way as they did.

One of the most impressive things about the Ju/wasi was their approach to Old Rule One. This involved their social system and knowledge of the natural world. Old Rule One is often translated as "survival of the fittest," and if we are social Darwinists, when we apply this to human beings we frequently have in mind such things as personal strength, individual abilities, and readiness for combat. Fitness, of course, involves many things such as intelligence, good eyesight, or resistance to disease, but in human beings, we seem to believe that it's a type of behavior, sometimes expressed as "might makes right" or "the good guy gets there last." But nothing could be further from the truth, as the Ju/wasi clearly demonstrated. Human beings are a social species. For our first 150,000 years on the savanna, we had to support one another, and the Ju/wasi continued this practice. Their population density was one person in 10 square miles, and thus they lived in fairly small groups spread across a dry landscape, with each group centered at a water source. Water, food, and protection from predators were of utmost importance, and to this end cooperation was essential. A person alone on the savanna could easily fall victim to a predator. Extremely young or old people could not get adequate food without the help of the strong adults. The young people were the future of the group, and the older people were its memory. In the case of a 50-year drought, for instance, the old people would remember how their group had survived the last one. If just one water source dried up, the people who depended on it would need to move to another source, which would have occupants of its own. The people in need would want to be welcomed there, not turned away. They would need the goodwill of the occupants.

Ju/wa culture contained multiple, redundant ways of cementing personal relationships and keeping the many small groups united. A complex kinship system, strong affinal relationships, and carefully tended gift partnerships were but three of many examples. The people had no chiefs or headmen. Everyone was equal. Everyone shared. Ill will and jealousy were considered to be dangerous, and the people did everything they could to avoid these threats to their unity. The "fittest" people were those who felt the importance of this most strongly. The "unfit" people were quarrelsome or aroused jealousy, thereby making themselves un-welcome. The really antisocial people, such as a man who once shot two other people with poison arrows, were sometimes killed—hunted down and executed. Thus the genes of the unfit, highly antisocial people didn't always show up in subsequent generations.

The Ju/wasi named themselves for their social excellence. *Ju* means person, *si* makes the word plural, and /*wa* can be translated as "pure," as clean water is pure. If someone approaches you without a weapon and says, "Mi /wa," he means "I'm harmless." Clearly, the social Darwinists with their notions of individual power and brute force have no concept of what survival of the fittest once meant to our species. The Ju/wasi knew very well, however. They were, evidently, the First People, and thanks to them and their social excellence, the rest of us are here.

The other aspect of Old Rule One—a knowledge of the environment—was equally critical, in that these hunter-gatherers had to know as much as possible about the land in which they lived. They had named virtually every species of animal and plant in the Kalahari, and knew everything that could be known about them that didn't require technical equipment. This was an enormous body of knowledge, and unlike us, who relegate the various aspects of such knowledge to certain specialists, it was held by everyone, even young teenagers, who through their childhoods learned from the adults and remembered.

Unfortunately for me in later years, I couldn't help but contrast the Ju/wasi with people of my own culture. One January afternoon, I felt unfair hostility to a student in our local high school angry at birds for not using their birdhouses. The idea that such birds survive the winter by migrating, not by staying in their houses, seemed unfamiliar to her. I also took excep-tion to a few high school students who, they said, were studying ecology and trying to give the town a decorative recycling bin. I'm not sure why they wanted to involve the town, or why they didn't do this for their school, as the town had been recycling for years but the school did not. Anyway, since I was involved with the town's side of the arrangement, I asked one of the students what aspect of ecology they were studying. He said they studied "the stuff around here." I asked, "What in particular around here?" He replied, "Oh, I don't know. Ducks, I guess." If this seems discouraging, try asking people where the sun rises. An astonishing number have no clue.

I often wonder what the above-mentioned high school students would think if they knew that teenagers like themselves who lived in the bush wore the skins of animals, slept on the ground, had never seen a town, vehicle, building, or even piece of paper, and certainly couldn't read or write, but knew infinitely more about the life of their ecosystem than all the above above-mentioned ecology students and their teacher put together.

Charles Handley, a field biologist who later became curator of mammals at the Smithsonian, came with us on one of our expeditions to the Ju/wasi and was substantially helped by a young Ju/wa man who was about the same age as these high school students. Charlie decided to test the young man's knowledge, and found he could name virtually all the different kinds of mice, shrews, bats, and other small mammals, to say nothing of the big ones, and also describe their habits. The youth knew birds and snakes too, plus insects and plants, but what interested Charlie were mammals. Charlie said that on almost all points, this teenager's information agreed with that of Western biologists.

A later visitor, a professor from Harvard with no connection to us, quizzed the Bushmen of another group about the natural world and concluded that "they knew almost as much as we do." But surely, they knew more. Since even now, the Western world does not know everything about the ecology of the Kalahari, the professor could only ask questions about things that were familiar to him, not those he'd never heard of. I think of a weaver bird's nest, which has an upper and lower compartment. The eggs or nestlings are in the upper compartment, but nothing is in the lower compartment, which the Bushmen said was for a snake. That sounds like superstition, yet a scientist was to learn that when a snake climbs on the nest, his weight opens the lower compartment as it closes the upper one, hiding the eggs or nestlings. Thus the snake assumes that the nest is empty and moves on. The natural world is full of things like that, some known, and some surely as yet undiscovered, but how could the professor ask about them if he didn't know about them himself?

To me, the greatest testimony to the environmental knowledge of the Bushmen was their arrow poison. Somewhere in the deep past, they discovered one of the deadliest poisons in the world. It comes from the pupae of *Diamphidia* and *Polyclada* beetles as well as their parasites, *Lebistina* beetles. A few drops of their poison will kill a twelve-hundred-pound African buffalo or two-thousand-pound giraffe. A single drop will kill a person. There is no antidote. How did people find this?

Everything mitigates against the discovery. There are thousands of kinds of beetles on the savanna, but only these are known to have poison. Some kinds of beetles are found almost everywhere, yet the poison ones are found only on a certain kind of tree that grows in sandy soil far from the seasonal lakes where people made their camps. To be sure, the trees are nut bearing, so the people certainly knew about them and would visit them when the nuts were ripe, and hence they would have noticed the beetles. But the adult beetles don't seem to be poisonous.

Some beetles such as scarabs interface with people in that they hang around their camps and clean up their latrines, but the poison beetles don't interface with people—they don't bite, or show an interest in our foods or feces, nor are they in any way conspicuous. *Diamphedia* beetles can be seen on the leaves of the tree (surely this is also true of *Polyclada* and *Lebistina*, although I didn't see them), but there's nothing eye-catching about them, and they're quite small. As for the larvae, although certain kinds of insect larvae are edible, the larvae of these beetles are tiny and never in view. The adults lay their eggs on the leaves of the tree, and when the eggs hatch, the larvae—sometimes with parasites attached—climb down the tree under the bark, exit the tree through its roots, and encase themselves with grains of the sandy soil until they look like little round sand balls. Inside these balls, three-feet underground, the larvae pupate. It's in this state that they're poisonous.

The early people, of course, knew about underground foods, which in itself was quite a discovery, and this knowledge greatly aided their survival on the savanna because few other creatures knew about edible roots or could dig down to get them. You can't do much better than discover an abundant, nourishing food supply for which there's little or no competition. But the roots that people ate were those of plants that grew aboveground, and no plants grew under the trees infested by the beetles except perhaps some grass. Maybe the trees, like black walnut trees, put out a toxin to keep other plants from growing nearby. Maybe the sandy soil was not conducive to a varied plant community. Maybe the lack of other plants has something to do with the poison beetles, or vice versa.

People would visit the trees to gather the nuts, but why would they dig around them when so obviously there was nothing there to eat? To get the pupae, a Ju/wa man might dig a hole about three feet deep with a three-foot diameter. The people had no shovels; they only had their hands and sharp sticks. Why dig so large a hole if one doesn't know what one is looking for? I know of no other item used by the Bushmen that is buried deep in otherwise-barren sand with no aboveground sign to show it's there. Old Rule One says not to spend energy and waste hard-won calories for no reason. Obviously the long-ago Bushmen did what some might have thought would be wasteful. Why?

And what about the pupa casings? Made entirely from sand, they are virtually invisible in sand, especially because all around them are almost-identical sandy lumps that are not pupa casings. Even though I often watched the holes that Ju/wa men were digging, I could seldom recognize the casings until the men picked them up.

Even so, long ago people not only found the casings but also wondered what was in them. It would seem that they opened them and found the pupae. Touching a pupa is safe enough, as the poison is inside the body, not out on the skin. One might think that someone ate a pupa and died, but probably that didn't happen. The Bushmen thought nothing of eating the blackened flesh around the arrow wounds of their victims—flesh that still contained poison—because the poison works by disturbing the blood and disintegrating the hemoglobin, and thus must enter the bloodstream directly.

Yet for some reason, someone extracted the poison. For one species of pupa (I'm not sure which), the Bushmen would tap it until its insides were blended, and then would pull off its head and squeeze a mush that looks like yellow mayonnaise on to the shaft of an arrow. Another species of pupa has a poison gland in its armpit, so to speak, and to get it, one pulls off a front leg and squeezes the poison, a drop of clear liquid, out through the wound. How could anyone chance on things like that?

We must assume that someone dug a hole under a tree where nothing grew, noticed a difference between the pupae casings and the other sandy round lumps, opened a casing, extracted the pupa, and for some reason examined its insides. Then what?

Perhaps someone had a cut on his hand, opened a grub, and somehow got the poison in the cut. Could that lead to the discovery? Not necessarily, because the poison takes so long to work. The person would die, but not for a day or two, during which time other distractions would surely intervene, so the cause and effect would not be obvious. The only suggestion I can offer is that perhaps the poison hurts. I once walked into the smoke of a fire in which Ju/wa men were burning the carcasses of the pupae after they had squeezed out the poison. The men yelled at me to get out of the smoke, and I did, although perhaps too late. I happened to have a small cut on my finger, and it soon began to hurt, then hurt a lot, and then hurt all the way up my arm. I can't imagine how the disintegration of one's blood cells can be painful, as I'd think the victims would just feel weaker and sicker until they died. But if the smoke wasn't dangerous, why did the men yell at me to get out of it? They knew. I can only guess. When I mentioned the pain, one of the men sucked my finger to remove any available poison, and here I am to tell the story.

As for how the poison was discovered, one could spin many different scenarios, but the real story will never be known. One thing is certain, however. Only people with the highest powers of observation and a breathtaking knowledge of their environment could make such a discovery. And how did they get this knowledge? Not as we might, after years of specialized formal education, and with the help of laboratories and scientific instruments, by instead by awareness of what was around them and the wisdom that anything perceptible, anything at all, could be important, even the armpit of a pupating beetle, one among thousands of other kinds of beetles, three feet deep in otherwise-barren sand. Such is the Old Way.

After I wrote a book about the Ju/wasi called *The Harmless People* (Thomas, 1959), I met a scornful graduate student in anthropology who said to me, "You seem to think those people were in harmony with nature." It was, at the time, politically incorrect to hold such a view, as the anthropologist Napoleon Chagnon had just written about Yanomama Indians in the Amazon jungles. Chagnon (1968) titled his book *Yanomamö: The Fierce People* (1968), perhaps to counter my title, although his title reflected his impression of the Yanomama, and mine was a loose translation of the word Ju/wasi, the people's name for themselves, and reflected their social values.

Yanomamö: The Fierce People showed that hunter-gatherers, or some of them anyway, were not interested in harmony. Well, the Ju/wasi weren't tree huggers like myself, to be sure—no one as enthusiastic about antelopes, hunts, and hunting, no one as ardent about tracking, the condition of one's arrows, or making a successful shot, could be called a tree hugger—but yes, I did think that they had "harmony with nature," or they did if any other species does, in that unlike myself and the graduate student, they were part of it all in the same way, say, as leopards, *Diamphidia* beetles, or black-eared, grass-climbing mice. No more and no less, as all of us, at one time, were part of it.

And unlike us, the Ju/wasi did not see themselves at the top of some taxonomic ladder. They said, for instance, that lions were better hunters than themselves, which was probably true, and is quite an admission, in that hunting was the most compelling, most all-consuming activity that these people knew. They belonged to a successful species, to be sure, but then, every other species on the savanna at that time was also successful in that they were all there, all living in the Old Way, all keeping the Old Rules.

In the 1970s and 1980s, due to the loss of much of their land and pressures from other cultures, the Bushmen were transformed from hunter-gatherers into farmers, laborers, clerks in government offices, or unemployed welfare recipients. Their lives, in other words, came to resemble those of other rural Namibians, which was their desire. This was the end of the Old Way as we knew it, and many of us mourn that fact. But after all, it's only gone for us and our domestic plants and animals. Everything else still lives by it, fortunately, as this is what keeps the planet going.

Even so, to experience the Old Way is important, not only for our personal pleasure, but also because it raises our awareness of the need for ecological responsibility. If we forget for a while our video games and television programs, our cell phones and Internet friends, if we get out of our cars and buildings, fascinating things are there for us to see. I've learned of a man who is writing a book about smoking. His habit takes him outdoors every time he wants a cigarette. As he smokes, he sees such fascinating things that he's writing a book to describe them.

As for me, if I saw the Old Way in Namibia, I see it again where I live in New Hampshire, not as a participant this time, but rather as an observer, and if it's not the Old Way of the African savanna that once involved our species, it's still the Old Way of the New England forests, and is just as interesting and significant in its own way. My husband and I live on land that my parents once farmed. My young grandson, Jasper, lives across the road with his parents. He owes his existence to my Old Way precautions, as the two-year-old boy who wasn't eaten by a leopard grew up to be his father. Often in the evening Jasper and I go out to look for "something interesting," as he puts it.

Sometimes we see whitetail deer in the field—usually a certain doe with her two grown daughters and her young fawn. We've seen them so frequently that they aren't just deer to us, they're individuals, all with their own identity and mannerisms. They know us too. They aren't particularly afraid but they also aren't happy to see us, so they move back into the woods. Old Rule One for the local whitetail deer: avoid people (Thomas, 2009). Sometimes we see a flock of wild turkeys in the field, getting ready to roost. Again Old Rule One, this time for turkeys: get to a safe place at night. Their timing seems perfect. They forage for food until just after sunset and then gather together in a pear-shaped flock at the edge of the woods to start the roosting process as the sky gets dark. They look at the trees for a while, no doubt selecting the branches they hope to roost on. Presently a turkey at the front of the group runs a short distance and launches into the air. Up she goes to the branch she has chosen. Then another turkey runs and launches, then another, then another, then two or three turkeys launch together, and then two or three more. The last turkey launches just when the night becomes really dark. If she waited any longer, she might have trouble finding her branch.

Jasper and I ponder these events. How do the turkeys decide who should go first? A flock of turkeys is a social unit, so some turkeys are more important than others. Does the most important one go first? And if so, is that one a leader or just more privileged? We wish we could answer these questions.

We notice other things too. On summer evenings we see clouds of dragonflies against the sun, weaving back and forth just above the long grass in the field. We know they are hunting, but we can't tell what. We find two infant garter snakes under a piece of bark and watch them push themselves quickly into the dust in an effort to hide. My dogs and I once encountered an infant rattlesnake who didn't hide but instead rattled at us. Thus hiding seems to be Old Rule One for small, nonpoisonous snakes. We cover them again with the bark.

Jasper is much taken with milkweed. We are able to observe the milkweed as its pods form, then as the pods open, and finally as the feathery seeds emerge from inside the pods and cling to the edges, waiting for the wind to carry them. Jasper has a collection of milkweed seeds, and ever since he was four years old has been able to recognize a milkweed at any stage of its development. How many kids, glued to their electronic devices, can say the same?

One afternoon we find a monarch caterpillar on one of the milkweeds and are able to observe his activities for two days. On the third day he disappears. This happens in the fall, and his is the generation that will migrate. Perhaps he has become a chrysalis and then a butterfly. We hope that his journey down the Atlantic coast will be successful. We have no idea how he knows where to go or how so a tiny creature could make such a journey. A monarch butterfly is not much bigger than a milkweed seed, which is designed so that wind will carry it. But the butterfly must resist the wind to stay on course. Having seen how the lightest breeze will toss a butterfly, we have no idea how he will manage. We only know that his journey is possible. We watch the weather carefully, hoping that this fragile little creature will not be blown out to sea.

All this and more is happening right outside our doors. We don't need to be scientists like those who fathomed the mysteries of Cordyceps, and we don't need microscopes or spaceships, as we would if we wanted to fully appreciate tardigrades, because every living thing has a seemingly miraculous story like none other. One year, for example, the oak trees throughout our region chose to eliminate some of their acorn-eating predators—turkeys, red squirrels, gray squirrels, chipmunks, black bears, moose, and whitetail deer, for instance; all these predators eat acorns in the fall in order to gain weight to carry them through the winter. Such relentless acorn eating spoils the oak trees' chance of reproduction. To put a stop to it, all the oaks of a region sometimes produce no acorns, thereby causing many of their predators to starve. In other words, the oaks were observing Old Rule Two, just as I was in Uganda when I withheld my children from a leopard.

The most amazing thing, however, is that all the oak trees of a region do this at the same time. I don't believe that anyone can predict when this will happen. This gives the act an intentional aura. Perhaps the oak trees send pheromones to each other. Other kinds of trees do this for other reasons, so it would seem that the oaks have some kind of system, as what good would it do a single oak to withhold acorns if the surrounding oaks were not doing the same? The withholding oak would just miss the chance to get its acorns planted by the local squirrels, and the other oaks would gain a reproductive advantage.

It's a good feeling to experience what's going on around you, whether or not you understand it. If you're fortunate enough to live where you can visit the same places frequently, you can keep track of the fortunes of the nonhumans there. For instance, the summer after the oak trees caused a near famine, they produced a huge crop of acorns, and we discovered a birdhouse packed with acorns right to the top. A squirrel had done this. He evidently had survived the earlier winter and wasn't going to let the oak trees deprive him again.

Also, a pair of bobolinks used to nest in our field, always in the same place. The male would perch on a certain, south-facing branch of a tree near the nest. The tree was in the middle of the field, the branch was about six feet off the ground, and the bobolink's sight line to the nest was at a 45° angle. He sat above the nest for long periods of time, and now and then would fly down to it. For two or three years he did this. Then one winter an ice storm demolished his tree. That spring I didn't see him, and I thought, because his special branch was gone, that he and his mate were nesting in someone else's field. But later that summer I saw him again, this time on a tree at the edge of our field. Again he was on a south-facing branch about six feet off the ground, and again his sight line to the nest was at a 45° angle. Why? There were lots of trees with lots of branches, but these bobolinks made their nests below the same kinds of branches at a distance that produced the same angle. I have no idea why, but surely they had a reason. If the Bushmen had observed something like this, they probably would know why.

So the bobolinks stayed, and I was happy. Many birds are endangered, especially ground-nesting birds such as bobolinks. This is because they nest in fields, and many fields are mowed in the spring, which destroys the nests and the eggs or nestlings. Some birds raise two or more sets of young each year, but not bobolinks, and that is why they are disappearing. Ours have remained because we don't mow the field until the fall—thus protecting not only the ground-nesting birds but also the fawns hiding in the long grass, monarch caterpillars in the milkweed, and many other creatures. Even so, some ground-nesting birds such as whip-poor-wills and killdeers disappeared from our area years ago, as birds world-wide are disappearing. If the endangerment of birds surprises you, please compare the dawn chorus of today to that which you heard years earlier, and please read "Emptying the Skies" by Jonathan Franzen (2010) in the *New Yorker* describing European birds and their problems. Europeans not only lack birds but also are actively killing those they still have. They wouldn't be doing this if they felt about them as I feel about the bobolinks. And I wouldn't have feelings for the bobolinks if I hadn't watched them for a while. Getting to know them didn't happen quickly and it wasn't entertainment, but it was real—very real. They were there.

Patience is the best tool for achieving this kind of knowledge. Our pace has been raised so high by our lifestyles that we need to work at being patient, but if we do, it pays. Some years ago, I was camped on Baffin Island watching a pack of denning wolves, and watched one wolf for eighteen hours without taking my eyes off her. This was during July in the high Arctic, so the sun didn't set, which is why I could watch for so long. The five adult wolves of that pack were raising a litter, and the wolf in question had just returned from a hunt. In denning season, wolves often hunt as individuals, not as a pack, and take turns staying near the pups in the den. Her duty, therefore, was to help feed the pack's seven pups, probably her younger siblings, who on her arrival came scrambling out of the den to beg for food. Her hunt had been successful. She lowered her head and vomited some chewed-up, partially digested meat. Within seconds, the seven pups had bolted down her offering and begged for more.

But the wolf didn't have any more, or if she did, she was saving it for herself. After all, even though she hadn't nursed the pups and thus could not have been their mother, she was still a responsible pack member who was helping to keep all of this going. She needed a little nourishment. So rather than deny the pups, she jumped up on a high boulder where they couldn't reach her. Discouraged, they soon went back in the den.

Exhausted from her hunting endeavors, the wolf fell asleep. I watched her. She slept for nine hours without moving. Then she opened her mouth just a little, shifted her position slightly, sighed, settled her tongue, and slept for nine more hours, again without moving. And all this time, I watched her. At last, another wolf came up to the den and fed the pups a little belch of something, perhaps a lemming. Most of the pups therefore got nothing, and they looked up at the newcomer, pleading. The newcomer didn't offer any more. The hungry pups cried. The sleeping wolf woke up, stood up, shook herself, and trudged off into the tundra to hunt and feed the pups again. I had been watching her for somewhat more than eighteen hours, during which time I came to understand her dedication to Old Rule Two and what it cost her. The observation took a little patience, yet it was worth the wait.

I don't remember what I did this morning, but I remember her.

References

Chagnon, N. (1968). *Yanomamö: The fierce people*. New York, NY: Holt, Rinehart, Winston.

Frazen, J. (2010). Emptying the skies. *New Yorker*, July 26, 48.

Margulis, L., & Schwartz, K. V. (1988). *Five kingdoms: An illustrated guide to the phyla of life on earth* (2nd ed.). New York, NY: W. H. Freeman and Company.

Thomas, E. M. (1959). *The harmless people*. London, UK: Secker and Warburg.

Thomas, E. M. (1965). *Warrior herdsmen*. New York, NY: Alfred A. Knopf.

Thomas, E. M. (1994). *The tribe of tiger: Cats and their culture*. New York, NY: Simon and Schuster.

Thomas, E. M. (2006). *The old way: A story of the first people*. New York, NY: Farrar, Straus and Giroux.

Thomas, E. M. (2009). *The hidden life of deer*. New York, NY: HarperCollins.

4

Wild Wings

Bridget Stutchbury

There is no doubt that people, young and old, need experiences with nature to feel and function well. Nature is restorative in terms of cognition and well-being, not simply because it is calming and peaceful, but also because it mildly stimulates our attention as well as replenishes our ability to direct attention to specific activities and problems. But what constitutes a nature experience? Studies have shown that merely seeing pictures of high-definition nature images does not have the same restorative effect as seeing, or being in, nature itself (Berman, Jonides, & Kaplan, 2008; Kahn et al., 2008; Kahn, Severson, & Ruckert, 2009). Marc Berman, John Jonides, and Stephen Kaplan (2008) found that performance on a demanding mental task improved when students took outdoor walks in natural versus urban environments. By extension, it is likely that a visit to the zoo or a neighborhood park, while beneficial, barely scratches the surface of what humans can experience, and need, from the wild. Humans might benefit most from in-depth, meaningful experiences with the wild rather than a domesticated version of nature.

We are losing our contact with the wild as a result of severe, ongoing biodiversity loss, and because so many people live in urban areas. The United Nations Secretariat of the Convention on Biological Diversity (2010) reports that since 1970, the abundance of vertebrate species has fallen by about 30 percent and that nearly a quarter of plant species are now threatened with extinction. In North America, Australia, and Britain, dozens of species of birds have declined steeply since the 1960s as a result of habitat loss and other threats (Stutchbury, 2007). Meanwhile human populations have been increasing exponentially, and today 82 percent of North Americans and 73 percent of Europeans live in urban areas (Population Division, 2007). The challenge for many of us, therefore, is finding the wild in nature.

This chapter considers what birds teach us about experiencing wildness and how the wildness of birds is being lost. Why birds? Bird-watching and backyard bird feeding are two of the most common ways that people experience nature. A US Fish and Wildlife Service (2009) survey found that 20 percent of adults in the United States consider themselves bird-watchers, which amounts to over forty million people. Most of these people feed and watch birds in their backyard, but half of them also travel away from home to see birds.

The key themes I develop here are that experiencing the wildness in birds involves going deeper in one's biological understanding and *not* just going farther afield to wild places; human-driven changes in the global landscape are domesticating wild birds, selecting for generalists over specialists, and shifting the baseline of what we consider wild; and many bird species are evolving in response to human changes in the global landscape, and therefore in an evolutionary sense we are losing some of the wild of birds.

The Wild Within

I have been a field biologist and ornithologist for twenty-five years, so you might think I have high standards when it comes to connecting with the wild and am satisfied only with sightings of hard-to-find birds. During the winter, at my farmhouse in rural Pennsylvania, I have a bird feeder on the porch only a few feet from the dining room window, and admire the nearly nonstop visits by ordinary birds such as hairy woodpeckers, downy woodpeckers, black-capped chickadees, tufted titmice, white-breasted nuthatches, slate-colored juncos, blue jays, and northern cardinals. These are some of the common feeder birds that I have seen thousands of times before, but every sighting brings an important token of a renewed connection with nature and, occasionally, a surprising observation. The experience of quietly viewing nature through a window, from the warmth and safety of your home, is obviously enhanced enormously by luring colorful and interesting birds to the scene with a concentrated source of food. Here, I argue that a deep knowledge and understanding of the natural history and behavioral ecology of birds provides us with a more meaningful experience than simply seeing birds as beautiful, entertaining ornaments for our backyards.

In eastern North America, chickadees are one of the most easy-to-identify visitors to bird feeders. Chickadees have a jet-black head and bib that contrasts with a bright white belly, and have a distinctive loud call ("chick-a-dee, dee, dee") for which they are named. The most casual observer may not even know the name of this distinctive bird (black-capped chickadee), yet surely will recognize it by color or sound. Since they can eat seeds, chickadees are among the few songbirds in northern regions that do not migrate south for the winter. They are entertaining to watch because they come in small flocks and are hyperactive, rarely sitting still for more than a few moments as they grab a sunflower seed and quickly retreat to a branch. As peaceful and interesting as this scene is, there is so much more to the lives of chickadees than we can see on the surface.

Chickadee winter flocks are comprised of breeding pairs who, during the summer, defended nesting sites and territories from each other. In fall, the breeding territory boundaries break down, and chickadee pairs come to together and cooperate to defend a much larger winter territory from neighboring flocks. That year's young birds, facing their first cold winter, join the adults. A chickadee flock is a complex social network and has a clear-cut pecking order, with older and larger birds in the top positions, and socially dominant over younger newcomers. There is a parallel hierarchy, one for males and one for females, and the top-ranked birds of each sex usually pair up and breed together the following spring. Social rank determines access to winter food and hence winter survival. Low-ranked birds are subordinate and promptly leave the feeder when a more dominant bird arrives, so as to avoid physical aggression and possible injury. Those low-ranked birds that do survive the winter often cannot breed because the top-ranked birds claim all the available nesting territories and there is not enough forest to go around for subordinate pairs.

Most young birds live as the lowest-ranked birds and wait in line for a higher social position, gradually moving up in rank in the winter flock as older birds die. Nevertheless, about a quarter of young chickadees take a different tack, drifting between flocks on a daily basis rather than having a single home flock. If a top-ranked chickadee disappears, then a wanderer quickly appears and jumps the queue, claiming the vacant position.

During the breeding season, the stakes shift from food and social status to sexual competition. According to DNA testing, although chickadees are socially paired and seemingly monogamous, many offspring are the result of females mating outside the pair bond. Female chickadees eavesdrop on male-to-male interactions by listening to dawn song contests between neighboring males that duel using their "fee-bay" song (Mennill, Ratcliffe, & Boag, 2002). A female can discern which male is dominant during boundary disputes, and then the next day she visits that neighbor to sneak copulations.

As I sit writing, watching the chickadee action at my feeder only a few yards away from my laptop, I don't just see cute, tuxedo-clad, bundles of energy. I see a social network of dominant and subordinate birds, an underworld strategy of flock switchers, and future philanderers. How can I not be awed by the world of the chickadee? Whether in your backyard or a wilderness area, knowing versus seeing birds is an entirely different experience with nature.

This can be thought of as a kind of nature literacy. If you pick up a book in a completely foreign language you may be able to enjoy the illustrations, but since you can't read the text you will get little out of the experience. By learning the natural and evolutionary history of birds, you become nature literate and can understand what you are seeing. Just as watching high-definition scenes of nature on television does not fully substitute for a window view of real nature, I would argue that viewing nature without knowing it does not fully satisfy the human need for other-than-human experiences. There is more to nature than attractive, peaceful scenery. There is a complexity and deep evolutionary history that teaches us what nature, and ourselves, are really all about.

Our ancestors had no understanding of evolution, prior to Charles Darwin's theory of natural selection, but they did have a comprehensive knowledge of the ecology of the plants and wildlife around them because it was a matter of survival. Today, many North Americans who live in suburban and urban areas are largely illiterate when it comes to nature. In Toronto, I have seen my neighbor's children scream in fear at the sight of a moth, and I know several parents who believe that the long-legged, harmless crane flies that hatch out of the lawn in late summer are in fact giant mosquitoes that must be dispatched at once with a tennis racquet or chemical attack. Yet I have met few people who are disinterested in learning more about the other-than-human world around them.

In the natural world, competition for sex and resources is near universal. Sex, adultery, divorce, and daily threats and acts of violence are common in the bird world of the backyard (Stutchbury, 2010). Male birds are under pressure to impress mates, females hold out for the highest-quality males, parents must share the burden of child care, and neighbors fight over space and food. The details of how these sexual and personal conflicts are resolved are the product of a long history of natural selection that favored winners over losers, gradually changing the genetic makeup of a population one generation at a time. An evolutionary lens can be used to better appreciate the melodies of the robin, flash of red on the cardinal, and amazing journeys of our songbirds that arrive so casually in spring after traveling thousands of kilometers.

Birdsong is music to our ears, but to birds it is also a sophisticated weapon to keep competitors at bay. Songbirds often have individually distinctive songs, and gauge their aggressive response to a singer according to prior social interactions with that bird and the possible threat it poses. Not every song is created equal, and some song types are saved for the most aggressive interactions. The trill, for instance, comprises almost-identical notes repeated in a fast succession that requires a precise coordination of vocal muscles and airflow. The sound produced is a trade-off between how quickly a bird can repeat the individual units versus the frequency range that each unit spans, such that rapid broadband trills indicate high male quality and signal that the singer should be treated as a serious challenge (Schmidt, Kunc, Amrhein, & Naguib, 2008). Females also listen to songs, and prefer males with more complex songs or a bigger repertoire. My research on hooded warblers has shown that males with a weak song performance have mates who sneak off territory to obtain extra-pair copulations from neighboring males that sing at a higher rate (Chiver, Stutchbury, & Morton, 2008). Most songbird neighborhoods are social networks where individual qualities are broadcast through song. A male territory owner recognizes his rivals individually, and continually updates his assessment of each rival's immediate and future threat, and meanwhile females are eavesdropping on male performances.

The stunning color patterns of songbirds are also used in communicating threat and individual quality. The common yellowthroat is a little warbler, and males have a Zorro-style black mask and challenge rivals with a ringing "witchity-witchity-witchity" song. The mask is used as a status signal, and males with bigger masks are socially dominant over other males as well as preferred as mates by females (Tarof, Dunn, & Whittingham, 2005). In some birds, like northern cardinals and house finches, it is the red coloration that reveals so much about the wearer (Hill, 2002). The intensity of red and orange colors indicates a bird's ability to find and consume foods rich in carotenoids. Eating carotenoid-rich foods does not guarantee a male will have sexy colors, because these chemicals are also used in the immune system. Females prefer red males because only a healthy male can afford the luxury of showing off carotenoids in his feathers.

People who provide nesting boxes for purple martin colonies spend many a lazy evening admiring the acrobatics of their tenants. There is no doubt that purple martins are masters of the sky, plucking dragonflies out of the air with ease and then flying into the impossibly small hole of their nesting compartment at breakneck speed. The martin's long tapered wings and streamlined body say it all: flying machine. The past few years I have collaborated with the Purple Martin Conservation Association to track martins to their wintering grounds in South America—the first time this has ever been done with songbirds (Stutchbury et al., 2009). Martins are described in the scientific literature as leisurely migrants, but our birds from northern Pennsylvania typically fly to the Gulf Coast states, across the Gulf of Mexico, and arrive at the Yucatán Peninsula within five days of leaving Pennsylvania. This is a trip of 2,400 kilometers, including an 800-kilometer overwater flight, in less than a week. In spring, most return from the Amazon basin of Brazil (7,500 kilometers) in only three weeks, averaging 350 kilometers per day. When viewing an iridescent black-blue male advertising his nest site with his complex, gurgling song or a female forcibly evicting another who dared to enter her nest cavity, it is hard to believe these birds have just flown from South America. Aristotle was so puzzled by the sudden appearance of swallows in spring that he assumed they buried themselves in the mud to survive winter, much as turtles and frogs do.

Nature literacy not only enriches our own experience of nature, it is also a powerful weapon for conservation and environmental sustainability. Margaret Morse Nice (1979, p. 43), a pioneering ornithologist famous for her bird behavior studies in the 1930s, wrote in her autobiography: "I thought of my friends who never take walks . . . 'for there was nothing to see.'" I was amazed and grieved at their blindness. I longed to open their eyes to the wonders around them; to persuade people to love and cherish nature.

She argued that the more we know and understand nature, the more we will care about what is being lost. Although many people do not have the time or means to travel to wild places, the songs, colors, and behaviors of common birds have much to teach us about the wild. The wild in this case is not one of remoteness or exotic scenery but rather of evolutionary adaptation that took place long before humans dominated the planet.

Wild Places

Though wildness can be found in backyards and small forest patches by appreciating the adaptations of birds, there are few among us who would not experience a greater sense of wildness and connection with nature when also in a wild place. Observing birds away from backyards, birders, and barking dogs gives me hope that I am witnessing nature on its own terms, free from human interference. It is almost as though I am in a time machine, traveling back to a period when birds went about their daily lives without facing human-caused habitat destruction, pollution, and biodiversity loss. There are also more surprises, dangers, and new experiences to be found in less familiar, and less traveled, places.

For many years, my husband, Gene, and I spent every winter in Panama, and took our young children with us. We lived in Gamboa, a tropical oasis where the highway ends and the town is surrounded by the Soberania National Park, Chagres River, and Panama Canal. Gamboa lies within the former Canal Zone, and the homes were originally built for the US citizens who lived there to operate the canal. Our son, Douglas, learned how to walk in Panama, and many of his first steps intercepted the paths of leaf-cutter ants walking along well-worn trails to their underground nests. Gamboa is unlike a typical chaotic Latin American village, and has neatly cared for lawns along with palm trees lining the streets, and is home to many biologists, like us, working at the Smithsonian Tropical Research Institute.

We were studying the dusky antbird, a small bird that skulks in the thick undergrowth and is more often heard than seen. On our annual visits to Panama, our job was to catch antbirds and equip them with tiny colorful leg bands, so that we could tell which antbirds lived on which territory and how long they lived. Dusky antbirds have a staccato song that increases in tempo and pitch ("da, da, da, da-da-da-da!"), and a pair frequently sings in a coordinated duet. Males and females are dead serious about defending their piece of the forest, and will come charging on attack toward other dusky antbirds that dare to intrude on their territory. We lured birds out of the undergrowth by playing a recording of dusky antbird songs to simulate an interloper. The territory owners usually arrived at the playback speaker within a minute and silently flew over the speaker looking for the intruder, mounting a surprise attack. Males are dark gray, and females are gray brown, and if angry they would fluff up their back feathers and hunch over, revealing a bright-white back spot. After a few minutes of searching unsuccessfully, pairs would begin singing to drive away the invisible threat.

The playback allowed us to see the birds up close with our binoculars and read the leg band color combinations of the old, familiar survivors on each territory. To catch unbanded newcomers we set up a special "mist net," named for its wall of fine nylon threads, by the speaker that was six meters long and two meters high, and held up with long poles at either end that were pushed into the ground. A mist net is rather like a large, rectangular, nonsticky spider web and difficult for the birds to see in the shady undergrowth of the forest. We surveyed almost all the dusky antbirds each year that lived along Pipeline Road, and our record for longevity was a male (red-red) who lived for ten years.

The Pipeline Road, a deeply rutted dirt road, penetrates far into the Soberania National Park and was built to establish a safe oil supply during World War II. Many of our subjects lived near the beginning of the road, in forest patches between the wide grassy navigation cuts that stretched from the hilltops down to the canal. It was quite surreal to be entering a lush tropical forest with our playback and netting gear, listening to howler monkeys roar in the distance, only to look behind us and see a massive container ship passing by on the canal. Doing a census of the entire dusky antbird population meant working our way down Pipeline Road while stopping the car to look for banded birds. The farther one gets into the forest, however, the narrower, bumpier, and less passable the road, even in a four-wheel drive jeep, and the more precarious the bridges over the streams. Our final census point, about six kilometers down the road, was most certainly a place where tropical birds live out their lives with little or no interaction with humans.

It is the amazing diversity of life and unique adaptations that draw researchers and birders to the tropics. There are about 970 species of birds in Panama—about the same number that can be found in all of North America. Many bird-watchers visit the tropics to see exotic new species, and are probably just as happy seeing a honeycreeper or motmot at a tropical bird feeder while having breakfast at the ecoresort compared to on a hot, buggy trail later that morning. But there are species and experiences that can only be found the hard way, by traveling into a wild place. A wild experience is enhanced by feeling privileged to witness a fascinating or rare event in a wild place, usually with few other witnesses, as opposed to seeing nature in a safe, comfortable, and easy-to-attain location along with other tourists.

One morning we had been doing a playback to dusky antbirds, and about a hundred meters down the road we saw the rapid movements of dozens of birds. As we approached we saw an army ant swarm; hundreds of thousands of ants covered the forest floor like a living carpet. When on a raid, the ants swarm over the forest floor like an amber tide that leaves few places for their victims to hide. At the leading edge of the swarm the ants kill millipedes, katydids, cockroaches, frogs, lizards, baby birds, and everything else they can overpower. I saw dozens of birds crowding around the army ants parasitizing their efforts, including the professional ant followers like the bicolored antbird, spotted antbird, and ocellated antbird. The antbirds perched within arm's reach of me, snatching up the insects that were fleeing from the ants. It was stunning to see the enormous numbers of ants that covered every branch and leaf within three feet of the ground, blanketing an area the size of a large living room, all frantically scurrying along but somehow well organized, with the swarm moving through the forest decisively and with well-defined boundaries. With a front-row seat to this spectacular nature scene, I felt invisible, but also in awe of the world of the antbird.

During our fieldwork, we were often treated to other flocks of birds that had gathered at food sources and grouped together for safety. Tropical gnatcatchers, forest *Elaenias*, and white-winged tanagers feed close to, and follow, the lesser greenlet, a small plain-looking bird that on first glance makes an unlikely flock leader. Up in the canopy we frequently saw troops of white-faced monkeys, crashing from branch to branch. Once, perched near the monkeys, I got my first-ever look at a double-toothed kite. This raptor is somewhat like the professional antbirds, but it follows monkey troops, not ants, and catches the insects and lizards that are stirred up by the monkey's crashing around in the treetops. Nearby was a flock of birds feeding at a fruiting *Miconia* tree. Golden-masked tanagers, plain tanagers, and red-legged honeycreepers were helping themselves to the all-you-can-eat buffet. A troop of Geoffrey's marmosets, colorful little primates the size of a house cat, came tearing by, barely pausing long enough to scold us.

The diversity of plants, birds, and other animals is at its highest in the tropics, and there is a bewildering array of adaptations to this unique and rich environment. Birds that follow ants, hawks that follow monkeys, ants that are fed by trees, wasps that live their entire lives inside a fig fruit, gigantic damselflies that look like living fossils from the dinosaur era, giant caterpillars painted black and orange to warn off any animals stupid enough to make an attack, and spiders that look like a bird's droppings; this is a biologist's paradise.

An important element of experiencing wildness is to feel vulnerable. In my Pennsylvania forest, I cannot recall in twenty years of research ever being in danger. I have come across an occasional black bear, but in each case the animal turned and ran away. The only time I felt truly scared was when a wild turkey suddenly sprang from the undergrowth and charged at me, screaming wildly and flailing its wings, to warn me away from her brood of chicks. Though I was in no danger from this bluff, the surprise attack was enough to spark brief but real panic (followed soon after by laughter on my part).

In Panama, however, the danger is real, and venomous creatures appear to be lurking everywhere. "Don't touch anything!" seemed the most sage advice. The *Paraponera* ant, or bullet ant, measures about two centimeters in length, and is very territorial, attacking any moving thing that comes its way. Gene has had the misfortune of being bit while walking through the forest and was laid up for two days in agony with an arm that swelled beyond belief. While in the tropical forest, we sometimes heard a loud buzzing sound overhead that grew to a deafening roar . . . then faded away in the distance. These were killer bee swarms on the move, looking for a new home, but fairly harmless since they did not yet have a hive to defend. Still, it was intimidating to once find thousands of bees parked at eye level at the side of the road—a seething mass of venom the size of a basketball. While working out of a small outboard boat on Lake Gatún, putting up bird boxes, we once bumped into a tree snag that happened to contain a killer bee nest, sending thousands of angry bees swarming out of the top like a mass of swirling smoke. We made a heart-pounding retreat, remembering fisherman on the lake who had been killed by a bee attack a few years earlier.

We spent days at a time living in Gamboa, immersed in the lowland tropical forest of Panama, but were not far from civilization. As one leaves town, the main highway to Colón on the Atlantic coast passes through Soberania National Park. One can occasionally spot birds like a slaty-tailed trogon perched on a vine over the road or a huge spectacled owl staring back at you from the edge of the forest with its hauntingly large, yellow eyes. But without warning, the forest is suddenly gone, and the trees are replaced with mile after mile of open pasture, small houses, stores, and bus stops. Impromptu piles of household garbage and wrecked cars are scattered along the roadside, and at dusty bus stops there are well-dressed ladies going off to work and groups of children in their neat school uniforms. Pastures are charred black right up to the road edge, and still smoking in a few places—evidence of the previous day's burning. If you look far to the left, as the highway crests a hill, you can peer through the haze and barely make out a forest along a distant ridge. That is the forested Canal Zone where we do our research, a narrow strip of forest that was preserved all along the length of the Panama Canal to prevent soil erosion from filling up the canal. Everywhere else, any forest without some kind of government protection has been cut down.

Dusky antbirds, double-toothed kites, and lesser greenlets that live within the national park are largely unaffected by the vast twentieth-century landscape that surrounds them. As nonmigratory birds, individuals rarely move more than a few kilometers from their birthplace and are unlikely to wander beyond the forest edge. The national park is large enough that it can sustain populations of truly wild birds, and give visitors and biologists a glimpse of highly specialized species that go about their daily lives in privacy. Yet as important as they are for protecting biodiversity and wilderness, large protected areas are scarce or nonexistent in many tropical countries.

Since the 1980s, Latin American countries have been clearing about 4 million hectares of forest *per year*, and this has claimed roughly 300 million hectares of tropical forest, an area about a third of the size of Canada or the United States (Williams, 2003). Latin America and the Caribbean currently are home to over 560 million people, and this figure is projected to climb to 710 million by 2030. This surge in the human population will put even more pressure on natural resources because people need homes, food, and a livelihood.

Complete deforestation is devastating for forest animals, and even when some forest patches are left intact the remaining populations are under siege. The ever-widening gaps between forest patches make it increasingly difficult for individuals to move between forest patches, isolating small populations in their respective fragments. Army ants need large home ranges to feed their huge colonies and disappear quickly from small forest fragments. So do the professional ant-following birds that lose their meal. In the central Amazon region of Brazil, for instance, 100-hectare patches of tropical forest lost half of their understory bird species within fifteen years (Ferraz et al., 2003). When a forest is lost, we do not just lose the plants and animals that once made up the complex community. We also lose history. Alive and buffered from modern civilization, these species give us a glimpse into the distant past and allow us to unravel the mysteries of life on Earth.

Shifting Baselines

As satisfying and effective as bird feeders and backyards can be for experiencing nature, viewing nature at home skews our experience of the wild by imposing a biological filter on nature. Backyard birds are those that prefer forest edges, shrubs, and open areas, and that can tolerate human activity. In ten minutes of idle watching, with cup of coffee in hand, I can easily count a dozen different species of birds using our farmhouse feeder, adding up to probably twenty or more individuals. In contrast, a cross-country ski trip through our 150-hectare forest in winter would, even after an hour, turn up only a few chickadees, a red-tailed hawk circling overhead, and perhaps the distant raucous call of a pileated woodpecker. Yet I would not be disappointed with this low bird abundance within the forest because it was precisely what I would expect; natural sources of food are scarce and spread out, so bird sightings and sounds in winter are few and far between. Bird feeders allow us to see that subset of birds that are willing to leave the forest, and create an artificial setting of frequent and high-intensity aggression over valuable, lifesaving food.

The tables are turned in spring, and the abundance and diversity of birds in the fields and forests surrounding my farmhouse is far higher than in the backyard. There are dozens of species of birds that can only be seen and appreciated by taking a short walk into the forest where they live. These are the forest specialists, mostly migrants, and they pour into the forest in spring to set up territories, vie for mates, and tend nests and young. Though not necessarily easy to see, each species has a unique song that can be learned to the point that recognition is automatic, just as we learn the words of a new language. I can stand in one spot, eyes closed with the damp rich smell of the forest around me, and in a few minutes reel off a long attendance list: two scarlet tanagers, four hooded warblers, four red-eyed vireos, an acadian flycatcher, two wood thrushes, two blue-headed vireos, an American redstart, a Blackburnian warbler, an ovenbird, and a rose-breasted grosbeak. At that moment, the fact that a morning hike down Pipeline Road in Panama would turn up dozens of different species is not especially relevant.

What I expect to see in my backyard or forest, at a given time of year and place, constitutes my baseline. Through my own experience with nature I have developed a sense of what to expect, or what is normal. I also experience this shifting baseline as I move from country to city because I expect far less from the feeder in my tiny backyard in suburban Toronto. There are dozens of homes on our street and precious few trees, and even in winter our feeder attracts mainly goldfinches and juncos. Chickadees and cardinals are cause for excitement, and the appearance of a hairy woodpecker or tufted titmouse would evoke astonishment and incredulity. I adapt to the relative scarcity of nature in Toronto, where the landscape is choked with homes, shopping centers, roads, and highways, by lowering my expectations such that even a goldfinch is a welcome sight.

Shifting baselines occur on a global scale too, and over a longer time scale. Birds that have become common in suburban and urban environments have a pre-existing flexibility to thrive in a wide range of environments, predisposing them to be able to live in open, human-dominated landscapes (Bonier, Martin, & Wingfield, 2007). This flexibility includes a behavioral tendency to explore and use novel habitats, food types, or nest sites, and a physiological flexibility to reduce stress levels in urban areas. As humans dominate and alter the landscape to our own purposes, we are selecting for bird species that are generalists, and also tolerant of human presence and disruption. The avian "winners" are those that are pre-adapted to what humans are doing to the planet, whereas the losers in this new evolutionary challenge are the specialists that are gradually or quickly disappearing. Highly sensitive species, with specialized lifestyles and few options, may simply go extinct.

The female robin that crouches on her bulky nest in the rhododendron beside our farmhouse has to put up with noisy kids racing down the driveway many times a day, barking dogs, slamming car doors, and feral cats on their nightly prowls. The male robins that bring music to the predawn darkness of our Toronto suburban neighborhood sing from rooftops, not trees. Robins in the United States have always been generalists, and nest in a wide range of habitats in nature; they evolved a built-in flexibility that today happens to allow them to breed in backyards. Other thrushes, like the wood thrush, do not have this flexibility and require a relatively untouched forest habitat. Although the wood thrush song is conspicuous and bold, these birds are shy and difficult to see. The construction of houses beside forests or the use of all-terrain vehicles within a forest can lead to wood thrushes abandoning the area altogether even if no trees are actually cut down. Wood thrushes have declined in number by about 30 percent since the mid-1960s (Sauer, Hines, & Fallon, 2008).

The extinction of the passenger pigeon, a forest bird, seems like a most unlikely event because less than two hundred years ago their population numbered in the billions. These pigeons formed enormous breeding colonies in eastern North America that stretched for dozens of kilometers, with ten or more nests per tree. The colony moved from one year to the next, flying thousands of kilometers searching for new forested areas where the beeches and oaks had produced a bumper crop of seeds. Because they were not tied to any one spot, pigeons always could find enough food to fuel their astounding numbers. The eastern deciduous forests of North America were heavily logged during the 1800s, but many small forest patches persisted through the worst of times. Why is it that forest thrushes, like the robin and wood thrush, survived this habitat destruction while the highly mobile passenger pigeons did not? It was the supercolonial behavior of pigeons that doomed them once the forests were cleared. Nesting in enormous colonies was a behavior as inflexible as their dependence on the seeds of forest trees. Small forest patches did not allow for large colonies, and unfortunately the birds refused to breed in small numbers. Deforestation happened so quickly that it was impossible for the pigeons to adapt their specialized behavior to the fragmented forests that had suddenly become the normal environment in eastern North America.

As hundreds of bird species become uncommon, rare, or even extinct, this diminishes our experience with the wild and resets what we consider normal. We, and our children, become satisfied with what is left, and do not necessarily even know what we are missing out on. The diversity and abundance of nature that a generation expects to experience (i.e., what is considered normal) is an outcome of their experiences in their youth. As the amount of environmental degradation and species loss increases on the time scale of decades, the baseline shifts each generation such that expectations match what is currently normal (Kahn et al., 2008).

Losing the Wild

The *Oxford English Dictionary* defines the word wild with respect to a place as "uninhabited, uncultivated, or inhospitable," and with respect to an animal or plant as "living or growing in the natural environment; not domesticated or cultivated." Though dusky antbirds in Panama are wild birds, it is reasonable to ask if the chickadees, nuthatches, and woodpeckers at my bird feeder are undomesticated. I would argue they are not truly wild because they have altered their natural behavior in response to a human activity and concentrate in unnaturally high numbers, reminiscent of pigs at a trough. Many times a day, I can watch the white-breasted nuthatch spread its wings and stubby tail, to make itself look bigger, and boldly charge the chickadees and titmice that land on the feeder. But I have not yet seen a nuthatch do so in the woods behind our house; such threat displays occur rarely because critical resources are not normally concentrated.

Each spring, in early May, we mix sugar and water, add the liquid to two feeders, and put the feeders out to attract ruby-throated hummingbirds. I can sit only five feet away and still watch females buzz to the feeder, maneuverings sideways and backward to lick up the nectar, then accelerate with lightning speed as they rocket to a nearby tree. A male lurking in the crab apple tree may give his courtship display by repeating a dozen rapid U-shaped arcs, somersaulting at the zenith of each flight. With supersize nectar sources, there is plenty of aggression as a half-dozen hummingbirds chase each other at breakneck speed in a futile attempt to monopolize the resource. This intensity of social interaction is not a normal part of hummingbird life; in the forest, females are solitary and raise young alone with no male help, and do not even live on a male's territory.

Bird feeders, backyards, and on a larger scale, human alternation of the global environment have changed the frequency, intensity, and consequences of individual behavior. Stephen Palumbi (2001) calls humans the world's greatest evolutionary force. Humans have changed the evolutionary playing field, creating new selection pressures as well as triggering short-term evolution through habitat destruction, overfishing, and climate change. This is a form of artificial selection, the process that humans have used deliberately to domesticate thousands of other plants and animals.

Backyard bird feeders in Britain have selected for the once-rare migration routes of European blackcaps such that off-course individuals now enjoy high winter survival and migrate earlier to the breeding grounds, getting a head start on their competitors. Populations of blackcaps in southern Germany and Austria that historically migrated to Portugal for the winter have, over the last thirty years, shown an increasing tendency to instead spend the winter in Britain (Bearhop et al., 2005). A survey of British backyard bird-watchers showed that in the 1960s, blackcaps were only occasionally seen during winter, but in recent years almost a third of backyard feeders were home to blackcaps. This shift in migration direction along with distance has a strong genetic basis, and the use of the new migration route has increased rapidly in just a few decades because of recent evolution. The high number of bird feeders in Britain, combined with climate change, has increased the winter survival of adults and may have made this new migration route possible by allowing those wandering individuals to return to breed. Individuals who inherit this gene migrate to Britain, just as their wayward parents did, but now have relatively high survival and reproductive success, and so the gene becomes ever more common each generation.

Songbirds in cities are evolving in response to the warmer environment that results from paved surfaces and buildings that retain and then radiate heat. Urban European blackbirds are more sedentary and breed earlier than their wilder counterparts in the forest. Jesko Partecke and Eberhard Gwinner (2007) tested whether these changes are the result of recent evolution by hand-rearing nestlings from Munich versus from a nearby forest. Males originally born to urban parents had a weaker migration activity and came into breeding condition several weeks earlier than birds hatched in a forest, indicating some genetic basis to these traits.

Thomas Smith and his colleagues (2008) found that African rain forest birds that live on cacao and coffee plantations, compared with the rain forest, have evolved longer wings (for longer flights in open habitat), duller coloration (to better hide from predators), and shorter songs (to be heard in the windier, hotter environment). In Vermont, about 40 percent of the hayfields are mowed early each summer, artificially changing the delicate interplay of male competition for mates and the mating system in a grassland bird, the savanna sparrow (Perlut et al., 2008). Normally the best territories are held by the largest males who attract multiple mates and also obtain extra-pair copulations. In mowed fields, larger males no longer held better territories, had only one social mate, and cuckoldry was rare. Although natural selection should favor males with a large body size, in mowed fields this was no longer the case.

Someone studying male starlings in southwestern Britain might find that female starlings prefer males with a high song rate, but that these males have weak immune systems. How could natural selection favor females that mate with males that have a poor ability to resist disease? The answer is that the behavior we would observe and measure today would not be a result of a long history of natural selection favoring advantageous mate choice but rather due to recent chemical pollution that disrupts mate choice. Starlings in southwestern Britain are regular visitors to sewage treatment plants where earthworms are laden with environmental pollutants, including one of the more notable feminizing chemicals, bisphenol A. Young starlings that were experimentally fed earthworms tainted with a realistic dose of this pollutant sang longer and had a repertoire size double that of males in the control group receiving untainted food (Markman et al., 2008). In mate-choice trials, females chose males with superior vocal performance because this strategy in the evolutionary past led to higher-quality offspring. But in this contaminated population, the choice backfires because male song is now a chemical indicator rather than a true test of male quality. Hormone-disrupting chemicals probably disturb birdsong in a wide range of species because these chemicals are used liberally around the world and are common contaminants in water supplies.

Birds give us a window into the distant past and help us to unravel the mysteries of life on Earth. The evolutionary forces that led to unique species and explain their complex adaptations are an irreplaceable piece of the history of our planet. Behaviors that we can see today allow us to reconstruct the past and understand how a process, repeated over hundreds of generations, could lead to what we see now. Painstaking observations of present-day traits along with carefully designed experiments can establish how aggression, cooperation, and communication affect an individual's survival as well as mating and reproductive success. From that, we can cautiously conclude that those behaviors that are successful today are the same that were successful in the distant past. Those behaviors that led to increased survival and production of offspring would have been favored by natural selection, and thus increase in frequency over time. The fundamental assumption of this evolutionary reconstruction is that contemporary processes are similar to the processes that occurred many generations ago.

Humans are causing rapid evolution of other species, not only interfering with our ability to reconstruct the natural world's past, but also diminishing the human experience of seeing a "wild" bird or other creature. One aspect of the wild is the sense of immersing oneself in other-than-human (i.e., not domesticated) nature and existing on nature's terms, even if only for a few hours, days, or weeks. In 1863, renowned English naturalist and explorer Henry Bates (1863, p. 53), who traveled the Amazon, wrote, "There is something in a tropical forest akin to the ocean in its effect on the mind. Man feels completely his insignificance, and the vastness of nature."

Nature is not so vast anymore, and humans are far from insignificant. Humans have an overwhelming presence on this planet, and our sheer numbers, the enormous scale of our footprint on the landscape, and the modernization of our agriculture, industry, and medicine means that we are now a key player in almost every ecosystem in the world. Whether through direct habitat destruction, or indirect pollution and climate change, we have dramatically altered habitats as diverse as coral reefs, tropical rain forests, and Arctic ice caps. Even birds that live in some of the most remote places on Earth experience the pressures of humans. A twenty-five-year-old albatross flying for thousands of kilometers over the ocean may never return to its breeding colony if it is snagged on the hooks of a longline fishing boat (Awkerman, Huyvaert, Mangel, Shigueto, & Anderson, 2006). King penguins nesting on Possession Island in the south Indian Ocean cannot find enough food to stay alive because ocean waters are too warm (Le Bohec et al., 2008). Quetzals, which are famous for their long emerald plumes, live at high elevations in rain forests, but are nevertheless exposed to high levels of the pesticide endosulfan, used widely in pineapple, banana, and coffee plantations, because it drifts into the cloud forest via prevailing trade winds (Daly, Lei, Teixeira, Muir, Castillo, & Wania, 2007).

The many examples of the contemporary evolution of birds in response to habitat change, climate change, and pollution include such a variety of traits (migration, exploratory behavior, song, color, shape, mate choice, and timing of breeding) that one could quite reasonably conclude that most bird species hold an anthropogenic signature within, and are becoming domesticated to some extent. In a sense, then, few species are fully wild anymore.

Conclusion

Birds are remarkably popular in our culture, giving tens of millions of people a much-needed link to the natural world. In our busy modern lives, a glimpse of a red-tailed hawk by the highway or chickadee at the feeder is a welcome sight, and can pique our imagination and curiosity. It is hard to watch birds for long and not be impressed by the complexity of their lives. Violence and aggression are easily witnessed, as are cooperation (real or imagined) among flock members and mates. I have been a guest on many call-in radio shows and there is no end to questions generated by backyard observations. One lady claimed to have seen a chickadee hit her kitchen window and fall stunned to the ground, only to be literally lifted away by the other chickadees in the flock. A common question is why does the robin, or sometimes the culprit is a cardinal, continuously peck on the windows of a house, apparently trying to get in. The answer is that they are not trying to enter the house but instead see their reflection in the window and are attacking the intruder. Why does the woodpecker rap so loudly on the stovepipe every morning at the crack of dawn? The answer is that woodpeckers drum on loud, resonating objects (usually tree trunks) to advertise ownership of the territory and warn off would-be intruders. Casual observations can inspire one to learn more about the natural history of common birds, and it is this nature literacy that leads to a deeper experience of birds and wildness.

A detective depends on evidence that has not been tampered with. The same is true for scientists who wish to understand how natural selection shaped the evolution of bird behavior. If someone were to bulldoze

Stonehenge, for instance, we would not just lose a puzzling array of large rocks but also lose our ability to figure out why the rocks were arranged in that particular pattern. Our children would never be able to visit this historic site and ponder the mysteries of the human mind. Even moving a single rock a few paces would make it difficult to test theories on the origin and design of this structure. The real-time genetic evolution of birds in response to the human domination of the natural world means that we are tampering with nature's evidence, and breaking the link between present and past forever (Caro & Sherman, 2011). Nature literacy includes an understanding of what is no longer wild in birds, and how what we observe could be an artifact of human changes to the natural world.

References

Awkerman, J. A., Huyvaert, K. P., Mangel, J., Shigueto, J. A., & Anderson, D. J. (2006). Incidental and intentional catch threatens Galapagos waved albatross. *Biological Conservation, 133*, 483–489.

Bates, H. W. (1863). *The naturalist on the river Amazons*. London, UK: John Murray.

Bearhop, S., Fiedler, W., Furness, R. W., Votier, S. C., Waldron, S., Newton, J., et al. (2005). Assortative mating as a mechanism for rapid evolution of a migratory divide. *Science, 310*, 502–504.

Berman, M. G., Jonides, J., & Kaplan, S. (2008). The cognitive benefits of interacting with nature. *Psychological Science, 19*, 1207–1212.

Bonier, F., Martin, P. R., & Wingfield, J. C. (2007). Urban birds have broader environmental tolerance. *Biology Letters, 3*, 670–673.

Caro, T., & Sherman, P. W. (2011). Endangered species and a threatened discipline: Behavioural ecology. *Trends in Ecology and Evolution, 26*, 111–118.

Chiver, I., Stutchbury, B.J.M., & Morton, E. S. (2008). Do male plumage and song characteristics influence female off-territory forays and paternity in the hooded warbler? *Behavioral Ecology and Sociobiology, 62*, 1981–1990.

Daly, G. L., Lei, Y. D., Teixeira, C., Muir, D.C.G., Castillo, L. E., & Wania, F. (2007). Accumulation of current-use pesticides in neotropical montane forests. *Environmental Science and Technology, 41*, 1118–1123.

Ferraz, G., Russell, G. J., Stouffer, P. C., Bierregaard, R. O., Pimm, S. L., & Lovejoy, T. E. (2003). Rates of species loss from Amazonian forest fragments. *Proceedings of the National Academy of Sciences of the United States of America, 100*, 14069–14073.

Hill, G. E. (2002). *A red bird in a brown bag: The function and evolution of ornamental plumage coloration in the house finch*. New York, NY: Oxford University Press.

Kahn, P. H., Jr., Friedman, B., Gill, B., Hugman, J., Severson, R. L., Freier, N. G., et al. (2008). A plasma display window? The shifting baseline problem in a technologically-mediated natural world. *Journal of Environmental Psychology*, *28*, 192–199.

Kahn, P. H., Jr., Severson, R. L., & Ruckert, J. H. (2009). The human relation with nature and technological nature. *Current Directions in Psychological Science*, *18*, 37–42.

Le Bohec, C., Durant, J. M., Gauthier-Clerc, M., Stenseth, N. C., Park, Y., Pradel, R., et al. (2008). King penguin population threatened by southern ocean warming. *Proceedings of the National Academy of Sciences of the United States of America*, *105*, 2493–2497.

Markman, S., Leitner, S., Catchpole, C., Barnsley, S., Müller, C. T., Pascoe, D., & Buchanan, K. L. (2008). Pollutants increase song complexity and the volume of the brain area HVC in a songbird. *PLoS ONE*, *3*(2), e1674. doi:10.1371/journal. pone.0001674.

Mennill, D. J., Ratcliffe, L. M., & Boag, P. T. (2002). Female eavesdropping on male song contests in songbirds. *Science*, *296*, 873.

Nice, M. M. (1979). *Research is a passion with me: Autobiography of a bird lover*. Toronto: Margaret Morse Nice Ornithological Club.

Palumbi, S. R. (2001). Humans as the world's greatest evolutionary force. *Science*, *293*, 1786–1790.

Partecke, J., & Gwinner, E. (2007). Increased sedentariness in European blackbirds following urbanization: A consequence of local adaptation? *Ecology*, *88*, 882–890.

Perlut, N. G., Freeman-Gallant, C. R., Strong, A. M., Donovan, T. M., Kilpatrick, C. W., & Zalik, N. J. (2008). Agricultural management affects evolutionary processes in a migratory songbird. *Molecular Ecology*, *17*, 1248–1255.

Population Division of the Department of Economic and Social Affairs of the United Nations Secretariat. (2007). *World population prospects: The 2006 revision* and *world urbanization prospects: The 2007 revision*. Retrieved from http://esa.un.org/unup.

Sauer, J. R., Hines, J. E., & Fallon, J. (2008). *The North American breeding bird survey, results, and analysis, 1966–2007: Version 5.15.2008*. Laurel, MD: US Geological Survey Patuxent Wildlife Research Center.

Schmidt, R., Kunc, H. P., Amrhein, V., & Naguib, M. (2008). Aggressive responses to broadband trills are related to subsequent pairing success in nightingales. *Behavioral Ecology*, *19*, 635–641.

Smith, T. B., Milá, B., Grether, G. F., Slabberkoorn, H., Sepil, I., Buermann, W., et al. (2008). Evolutionary consequences of human disturbance in a rainforest bird species from Central Africa. *Molecular Ecology*, *17*, 58–71.

Stutchbury, B. (2007). *Silence of the songbirds*. New York, NY: Walker and Company.

Stutchbury, B. (2010). *The private lives of birds*. New York, NY: Walker and Company.

Stutchbury, B.J.M., Tarof, S. A., Done, T., Gow, E., Kramer, P. M., Tautin, J., Fox, J. W., & Afanasyev, V. (2009). Tracking long-distance songbird migration using geolocators. *Science, 323*, 896.

Tarof, S. A., Dunn, P. O., & Whittingham, L. A. (2005). Dual functions of a melanin-based ornament in the common yellowthroat. *Proceedings of the Royal Society of Biological Sciences, 272*, 1121–1127.

United Nations Secretariat of the Convention on Biological Diversity. (2010). *Global biodiversity outlook 3: Executive summary*. Retrieved from http://www.cbd.int/doc/publications/gbo/gbo3-final-en.pdf.

US Fish and Wildlife Service (2009). *Birding in the United States: A demographic and economic analysis*. Addendum to the 2006 National Survey of Fishing, Hunting, and Wildlife—Associated Recreation. Report 2006–4.

Williams, M. (2003). *Deforesting the earth*. Chicago, IL: University of Chicago Press.

5

Children and Wild Animals

Gail F. Melson

Developmentalists are just beginning to consider the importance of wilderness and "wildness" in children's lives. This theoretical and empirical neglect of the "wild" is surprising, for a number of reasons. First, a contextual, systemic approach to the study of children's development (Fogel, Greenspan, King, Lickliter, Reygadas, Shanker, Toren, 2008; Melson, 2008) is now widely accepted. Since the publication of the groundbreaking classic *The Ecology of Human Development* by Urie Bronfenbrenner (1979), a study of child development in context—often called "the ecological systems approach"—has emerged as the dominant paradigm. This approach mandates careful attention to all elements—physical, social, and emotional—of a child's context. It is also evident that the contexts of development include many nonhuman life-forms, including animals. Third, beyond animal presence, children's interest in and involvement with other animal species (Melson, 2001), nonanimal life-forms such as plants, and natural environments are now well documented. This responsiveness to nature is consistent with the biophilia hypothesis (Kellert & Wilson, 1993; Kellert, 1997; Wilson, 1984), which argues that since humans coevolved with other animals and life-forms, humans are innately attuned to them as well as to aspects of natural settings associated with survival (e.g., savanna-like vistas affording shelter and visual inspection of the surroundings).

Nevertheless, children's interest in, ideas about, and engagement with wild animals has been largely ignored. One may speculate that this void is an adaptation to the widespread view that wild animals have been disappearing from the contexts of development for children in both industrialized and developing countries. If children's engagement with wild animals is perceived to be of historical interest only, its relevance to contemporary child development can be easily discounted. Further, in a process Peter Kahn (1999, p. 7) calls "environmental generational amnesia," people may take the natural environment they experience in childhood as the norm against which to measure later environmental changes. This process leads even scholars of child development to view the absence of the wild and wild animals in particular from children's lives as the norm, and not as evidence of an already-impoverished context. Finally, the supposed disappearance of the wild (along with the dwindling number of children who have direct contact with domestic farm animals) has thrown children's relationships with companion animals into sharper relief. Pets thus have been singled out for study as the only animals that share children's daily lives. While the developmental, educational, and therapeutic significance of the child–pet relationship is now well established (Fine, 2006; Melson, 2001), narrow focus on the importance of pets has led to a tendency to conflate child–animal interactions with child–pet interactions. In this way, dogs, cats, rabbits, guinea pigs, hamsters, fish, and turtles—the most common species kept as pets—have become, in the words of Paul Shepard (1995), "the others," ambassadors of the wild animal species no longer present.

This chapter challenges the widespread belief that wild animals play no role in children's lives. I argue that contemporary contexts of development, including those in urbanized, industrialized countries, still include wild animals. While the human-to-wild-animal connection has changed historically, it remains important. Nature, the wild of this book's title, continues to play a part in children's development, including in densely populated urban settings. Indeed, contemporary children's connection with wild animals and natural environments may be more complex as well as more crucial than ever.

The central question this chapter asks is: What is the significance of contact with wild animals for children's development? Before one can entertain answers, however, the question itself needs unpacking: What is meant by a wild animal? What does contact mean? As Stephen Kellert (2009) contends with respect to children's interactions with nature, scholars need to ask the descriptive "What?" questions—How, when, and where?—as well as the "So what?" questions—What are the developmental consequences? What aspects of children's development are relevant? What components of wild animal contact are important?

In this chapter, I draw on existing theory and research from a variety of disciplines, including psychology, sociology, animal science, environmental science, geography, and the emerging field of human–animal interaction studies. Despite the range of relevant disciplines, both theory and empirical research on the significance of wild animals for children's development remain limited, for the reasons noted above. Moreover, no coherent conceptual framework sets the major issues, outlines the research agenda, and draws implications for education, parenting, social policy, and urban planning. The chapter seeks to begin such an endeavor.

What Is a Wild Animal?

A conventional definition of a wild animal is a nonhuman animal species that is not domesticated. Within this broad definitional boundary lies a wide variety of species. There are wild animals, such as lions, tigers, rhinos, and bears, associated with specific wilderness habitats. Such large mammals are the species most frequently thought of as wild animals in contemporary Western popular culture. Many other wild animals, however—for example, birds, squirrels, foxes, rats, and mice—generally inhabit the edges or centers of human habitation, including dense urban areas, surviving opportunistically on human material resources (Palmer, 2003). Still other wild animals have wholly dependent relationships with humans because of intentional human behavior, such as wild bird feeding, falconry, carrier pigeon use, or the keeping of wild animals as pets (Drews, 2002). Wild animals in zoos, aquariums, and nature parks as well as research facilities exist in total dependence on humans within environments designed for specific purposes. In addition, feral animals, such as feral dogs and cats, are wild animals that once had been kept as pets, but are no longer serving as companions or workers for humans.

Many species typically inhabit multiple contexts and have diverse relationships with humans. Tropical fish, for instance, may be viewed as pets in a child's bedroom aquarium tank, but as wild animals in the Caribbean coral reefs that are their native habitat. Some rabbits are kept as pets, others are raised as production animals for food or fur, and still others are observed as wild animals scampering around the garden patch. Thus, rather than a sharp ontological divide between wild and tame, or wild and domesticated, I suggest, following Clare Palmer (2003), that there are gradations of animals' dependence on and connection with humans, with no clear division between the wild and not wild.

Moreover, contact with wild animals always occurs within a specific context, varying in its wildness. Like the concept of wild animal, the notion of wildness or the wild—treated in more detail elsewhere in this volume—is not a simple binary property of an environment. Human visitors to some areas, such as parks and nature preserves, may go for immersion in nature, yet these environments may have been dramatically transformed by human design in order to give the visitor the felt experience of wildness. In other areas, like the Kalahari Desert environment of the hunter-gatherer groups known as the !Kung (Thomas, 2006), the distinction between human habitat and the wild hardly makes sense. Human animals in this case are simply one element of an intricate ecology of various animal species and plants, all living in mutual dependence. All the life-forms of this ecology—human, other animals, and plants—are continually changing and modifying it.

Wild Animal as Historical-Cultural Construction

Wild animals exist in a range of environments that vary in the extent and nature of their dependence on humans, although all bear some mark of human imprint. Following the description by Erving Goffman (1963) of the social organization of public places, one may suggest that context sets the "frame" within which humans define an animal as wild or in the wild. In this sense, wilderness and the wildness of an animal are social constructions that vary culturally, historically, and idiosyncratically.

Historical changes in ideas about wild animals, children, and the relation between them illustrate this point. Over a wide swath of human history—for example, the last twenty thousand years—one might trace changes from hunter-gatherer views of animals as other sentient and co-equal beings, immediately and deeply entwined with human experience and survival, to the ideas of early agriculturalists, who developed a vested interest in demarcating human (and domestic animal) territory and resources apart from those of the wild animals surrounding them.

A more recent example of changing ideas comes from Europe and North America, from the nineteenth century to the present. The nineteenth-century view was that wild animals were savage threats to humans and had no rights, and therefore could be mistreated, killed, or indeed wiped out with impunity. Hunting, the extermination of species such as wolves as threats, and "animaltainment," such as bear pits or organ grinders with their monkeys dressed in outfits, were all celebrated in books and toys for children (Oswald, 1995; Varga, 2009). Wild animals were juxtaposed in contrast to "civilized humans," with the former completely subservient to the needs and wishes of the latter.

By the end of the nineteenth century, however, evolutionary ideas forced a rethinking of the supposed dichotomy of human and wild animal. Ideas about wilderness as the source of beauty, contemplation, and transcendence—not merely threat—were gaining in the nature writings of

Henry David Thoreau and John Muir, even as anticruelty and animal protection movements were growing. Notions about childhood as a period of "natural" innocence led to the view that children, free from the strictures of "civilization," shared an innate kinship and connection with animals, including wild animals. Stuffed animals, from the original teddy bear to cuddly versions of almost every wild animal, became ubiquitous (Varga, 2009). Children's books about wild animals shifted from the nineteenth-century themes of heroic killings of savage, threatening beasts, such as bears and wolves, to wild animals as hapless victims of now-savage humans (Oswald, 1995).

As a result of these historical shifts, attitudes toward wild animals changed from viewing them as a threat to civilization to viewing civilization itself as a threat to wild animals. Thus, according to recent survey research, most young adults (and females more than males) from the United States, Europe, and Asia consider zoo confinement and wild animal traps unacceptable because they are seen as harming wild animals. Among respondents in these countries, only those in the United States find sport fishing or hunting deer or fox for sport unobjectionable (Phillips & McCulloch, 2005).

Despite the ascendance of the theme of wild animals as "noble beasts" rather than savage threat, scholars conclude that the earlier discourse persists, albeit as a minor theme (Varga, 2009). This suggests that historical changes in social constructions of the wild never fully supplant earlier ideas but instead layer over them. Even the earliest human conceptions about wild animals, as evolved over millions of years as hunter-gatherers, may lie as a substrate within human minds as they think about wild animals.

Contexts of Wild Animal Contact

The contexts of wild animal contact for children vary along dimensions of immediacy, structure, and affect. Hunter-gatherer societies, like the Inuit of Alaska or !Kung of the Kalahari (Konner, 2005), resemble the environments of evolutionary adaptiveness throughout 99 percent of human history. In these societies, humans share ecological niches with the wild animals on which they depend for survival. These animals are central to hunter-gatherer morality, spirituality, and social cohesion (Nelson, 1993). Although contemporary hunter-gatherer societies are dwindling in number, and none are fully isolated from agricultural and industrialized groups, hunter-gatherer children continue to have immediate regular contact with the animals that are food and clothing sources for their culture. Their relationship to these wild animals is infused by the hunting traditions of their culture. Complex and highly charged emotion, often embedded in rituals of the hunt and its aftermath, surrounds the animals.

In urbanized, industrialized societies, less immediate, more narrowly focused contact with wild animals is more common. Many children observe and sometimes attempt to interact with wild animals in parks, nature preserves, zoos, aquariums, butterfly farms, bird sanctuaries, and the like. S. K. Turley (2001) estimates that children account for 37 percent of all visitors to such wildlife settings. Children may feed wild ducks at a park or handle (under adult supervision) a wild animal, such as a hawk or bat, as part of a nature education program (Sorge, 2008). Many families observe wild birds at backyard bird feeders, see squirrels, rabbits, frogs, and other creatures around the neighborhood, and discover that they share their homes unwillingly with ants, spiders, flies, cockroaches, and mice.

Such everyday encounters with wild animals illustrate the fact that outside of contexts designed to mediate children's direct encounters with wild animals—zoos, parks, and aquariums—many spontaneous observations of and interactions with wild animals persist in and around the child's home, school, and neighborhood. Squirrels nest in a backyard tree, bumblebees circle a picnic bench, earthworms wriggle in the dirt, and rabbits invade the family garden. Even in the most urbanized environment, children have contact with wild animals every day. Because cultural constructions of wild animal center on prototypes of large mammals in wilderness settings, these encounters with squirrels, bees, worms, rabbits, and the like are rarely thought of as *wild* animal experiences. Yet these experiences may be important developmentally.

Structured, adult-mediated environments for encountering wild animals inevitably narrow the child's range of experiences with the wild animals in question in a number of ways. First, adults frequently design and supervise the child's experience, encouraging certain behaviors (e.g., feeding the ducks) and discouraging others (e.g., chasing them to see them fly away). To take another example, a backyard bird feeder focuses the child's attention on bird behavior only at and around the feeder. Adult mediation further constrains the child's attention to certain aspects of the animal and its behavior. In observing birds at a feeder, for instance, adults might encourage the child to notice diacritical markings that aid in species identification. Second, settings structured by humans to provide a "nature experience" with wild animals also elicit a relatively narrow range of animal behaviors when compared with the animal's full repertoire of behavior in its ecological niche. Environmental demands may indeed so modify animal behaviors (e.g., ducks "begging" for bread crumbs brought in by human visitors to the park), that such behaviors bear little resemblance to naturally occurring patterns. Third, adult mediation of children's experiences with wild animals conveys cultural and moral messages concerning human–animal relations. To continue with the example of feeding ducks at a park, the experience may be framed as a way to bring animals normally wary of close human contact into touching distance and get them to do what we humans want. In the process, the insistent quacking, pushing, and competition for thrown bread crumbs may be framed as amusing, "silly" animal antics. Adult structure and mediation of children's engagement in this case with wild animals invariably result in relatively impoverished experiences perceptually, cognitively, and even morally.

The emotional impact of exposure to wild animals is another key dimension of the child's experience. One cannot assume that only immediate, unstructured encounters with wild animals engage the deepest feelings. Activities like feeding ducks at a park, watching squirrels in the backyard, or wild bird feeding may evoke high or low affect from children (and the adults with them). A child may be excited to see, for example, a relatively rare species of bird appear at the bird feeder and hence watch the bird intently. On the other hand, the child might pay little attention and show little emotion when noticing the regular gathering of birds around the backyard bird feeder. Again, adult mediation encourages certain emotional "scripts"—disgust at ants on the kitchen counter, awe at a hawk circling overhead, or laughter at the "tricks" on display at a dolphin show. The child's temperament and prior as well as concurrent experiences and relationships also undoubtedly play a role. While there is wide variability in children's emotional engagement with wild animals and wild places, we know little about what child, animal, or environmental factors are at play.

In addition to contextual characteristics such as immediacy, structure, adult mediation, and emotional tone, the frequency and duration of children's contact may be important. Just how much time does a child spend observing, interacting with, or being in proximity to a wild animal? In contrast to research on children's time spent with pets (Melson, 2001), there are no reliable estimates of the frequency or duration of children's attentiveness toward or contact with wild animals. While analyses of zoo and aquarium attendance suggest that children are disproportionately represented, these figures convey no information about the frequency or duration of focused attention on the animals, or on informational displays about them. Turley (2001) found, in a UK study, that family groups with children were most interested in "family togetherness" and "a fun activity" as motives for zoo visits, with interest in observing or learning about wild animals a lower priority. Children therefore engaged in relatively little observation of the animals themselves. Kellert (2002) argues that zoos, nature programs, and other structured experiences tend to be sporadic, focused on the rare and unusual, and lack intimacy, challenge, and creativity. He concludes that as a result, such contexts are impoverished sources of information, experience, and learning.

Contested Contexts: The Example of Zoos

The experience of zoo going serves as an illustration of the complexities of mediated encounters with wild animals. An estimated 98 percent of all Americans report having visited a zoo at least once (Dunlap & Kellert, 1994). A UK study of zoo behavior (Turley, 2001) found that adults organized zoo visits primarily for children, with family groups containing children ages five to twelve most likely to visit. Yet cross-cultural attitude surveys of young adults find considerable uneasiness with wild animals in zoo confinement (Phillips & McCulloch, 2005). Most respondents to surveys conducted by Phillips & McCulloch (2005) agreed with the statement that "many wild animals suffer considerably from stress and boredom, as a result of being kept in zoos," and disagreed with the claim that "the educational and entertainment value of zoos is far more important than any cruelty that may be involved in holding wild animals captive." (Children's views of wild animals in zoos have not been studied.)

Kellert (2002) argues that indirect contact with wild animals in zoos and nature programs is a poor substitute for unstructured encounters with wild animals in the context of exploration of natural environments. Stronger objections come from John Berger (1991, p. 16), who contends that the very act of observing animals in a zoo objectifies the animal and separates the human viewer from the subjectivity of the animal: "Animals [in zoos] are always the observed. The fact that they can observe us has lost all significance." On the other hand, Olin Myers and Carol Saunders (2002) note that when children observe zoo animals, or even better, can touch them at petting zoos, the children can see good models of animal welfare, conservation, and appropriate animal care. Seeing wild animals and learning about them in zoos can be a critical source of learning about other species. Berger (1991, p. 16), however, views even such learning with alarm: "They [wild animals] are objects of our ever-extending knowledge. What we know about them is an index of our power, and thus an index of what separates us from them. The more we know, the further away they are." The contexts of the zoo, aquarium, or nature park appear to be "contested frames," in which the meaning of encounters with wild animals is hotly debated.

Underlying these differing views may be contrasting stances on optimal contexts for children's learning. From a Vygotskian perspective, children's learning is optimized when more skilled individuals, usually adults, guide the child from what they already know to greater competence of predetermined outcomes, working within the child's "zone of proximal development" (Vygotsky, 1978). In this view, wild animal encounters that are mediated by a knowledgeable adult who is sensitive to the child's developmental level exemplify an optimal learning context. An alternate perspective sees learning as optimal when child directed, and based on exploration and discovery of places free from adult structure or supervision (Pyle, 1993). From this perspective, unmediated, unstructured direct experiences with wild animals would result in more cognitive and emotional growth.

Wild Animals through Media Imagery

The least immediate, most indirect forms of contact with wild animals are rapidly becoming the most common for children. Even those living in rural settings near wilderness areas such as national forests report that they see more animals on television and in movies than in the wild (Nabhan & Trimble, 1994). All children experience wild animals on the symbolic plane, in stories, toys, and other artifacts. This symbolic plane, what Gary Nabhan and Stephen Trimble (1994, p. 86) term "a vicarious view of nature," is now the most dominant form of engagement with wild animals and wilderness more broadly. Nature documentaries, cartoon wild animals, Animal Planet, and stuffed animals provide children highly scripted narratives of wild animals and wilderness. Background music and film editing help create emotional narratives, such as the "amusing antics" of wild animals, or "frightening" vision of one animal killing and dismembering its prey. Although representations of wild animals are proliferating in print and visual media as well as on toys, Kellert (2002) points out that animal symbols, reflected in cave paintings, totems, animal legends, and myths, are as old as the human species itself.

This symbolic plane casts certain wild animals—usually large carnivorous mammals—as more prototypically wild than others. Some animals—pandas, dolphins, and Emperor penguins, for instance—are presented as cute and friendly, while others, such as sharks and wolves, are portrayed as threats. In this way, some wild animals, encountered only through media and toy images, may become more salient for children as markers of emotion and personality than the living wild animals in their immediate environments. As an example, four- to five-year-old Maltese children identified tigers and lions, not found locally, as their favorite animals, and mentioned materials such as toys, clothing, and school stationery, as sources for encountering them (Tunnicliffe, Gatt, Agius, & Pizzuto, 2008).

Virtual reality is increasingly able to evoke feelings of immediacy, involvement, and emotion similar to those that actual encounters with wild animals produce. Emerging evidence from cognitive neuroscience (Decety & Jackson, 2004) shows that mirror neurons enable one to experience the emotions of direct engagement even when only observing or simulating it. These findings indicate that all forms of engagement with wild animals, even the seemingly most mediated and symbolic, may nonetheless be powerful experiences. At the same time, advances in virtual reality raise questions about what is lost when direct, unmediated engagement with wild animals and their environments disappears.

Children read about as well as look at stories, games, toys, and other artifacts about wild animals with surprisingly high frequency. As of 1998, seven of the top ten all-time best-selling children's books in the United States were about animals, both pets (*The Pokey Little Puppy*) and wild animals (*The Tale of Peter Rabbit*) (Melson, 2001). Three-quarters of a random sample of US children's books published between 1916 and 1950 had animal characters (Lystad, 1980). Nearly a third of the stories in fourth-grade school readers published in the United States from 1900 to 1970 have animal characters, half of which are the main protagonists (Melson, 2001). Moreover, children often prefer stories and other materials about animals to those about peers or other humans. In one study (Boyd & Mandler, 1955), when third graders heard stories with animal characters and identical stories with human characters, 75 percent of the children preferred the animal stories. Unfortunately, this study, along with the others cited above, failed to distinguish between wild animal characters and other types of animals. Given the pervasiveness of exposure to wild animal imagery and content in various media—books, toys, video, television, and Internet sites—it is surprising that so little is known about the impact on children's interests, ideas, or values concerning wild animals.

Just as zoos and other mediated forms of contact with wild animals evoke both positive and negative reactions, so too does the symbolization of wild animals through fiction, video, and other media. On the one hand, wild animal symbols are pervasive. They play a role in children's developing sense of self—an issue addressed later in this chapter. On the other hand, some scholars decry animal symbols as a demeaning, reductive process that reduces the animal to an instrument or resource for human needs. As an alternative to a symbolic role for the wild animal, Kenneth Shapiro and Marion Copeland (2005, p. 344) advocate that it appear "as an individual with some measure of autonomy, agency, voice, character, and as a member of a species with a nature that has certain typical capabilities and limitations."

The varied forms of contact, from immediate to mediated to symbolic, and the varied contexts in which they occur are likely to be interdependent. Cultural views of wild animals and wilderness, transmitted to children via symbolic materials such as stories and toys, may influence how children think about and respond to living wild animals (Varga, 2009). For young children particularly, compelling animal characters may obscure accurate understandings of wild animals. Every whale can become a Free Willy, every lion a king, and every vulture a plotting enemy. A humane educator reported (Melson, 2001) that children who had seen the animated movie *Pocahontas*, in which a hummingbird character attacked people, believed that the purpose of a hummingbird's long pointy beak was for pecking children's eyes out. Conversely, a child's direct observation of a wild animal, such as a rabbit, may help the child distinguish between its behavior and characteristics and those of Peter Rabbit. Children's exposure to animal symbols can provide teachable moments for adults to help children distinguish between accurate and symbolic, anthropomorphic animal characteristics.

Wild Animals and Children's Development

As Kellert (2002) notes, any experience with nature—direct, indirect, or vicarious (or symbolic)—may prompt cognitive, emotional, and moral learning, and have developmental consequences. With respect to engagement with wild animals specifically, little research documents links to developmental outcomes. Moreover, it is unclear how the varying modes of experience with wild animals described above differ in their developmental consequences. Some (Louv, 2006) fear that symbolic experiences of wild animals and nature lack the rich multisensory learning of more direct engagement, and hence lead to an impoverishment of experience. While provocative, this prediction currently lacks compelling evidence. To stimulate a research agenda in this area, I examine indirect evidence and theoretical support to raise questions about ways that wild animals may play a role in various developmental domains, specifically perceptual-cognitive, self, social, and moral development. In doing so, I suggest hypotheses about how differing modes of engagement may relate to children's development.

Perceptual-Cognitive Development
At least three areas of research and theory support the hypothesis that encounters with wild animals may promote perceptual-cognitive development. First, as noted earlier, the biophilia hypothesis (Wilson, 1984) predicts that children (and humans of other ages) will perceptually orient toward life-forms, including animals. This differential attention implies that living things will be preferential sources of perceptual and cognitive information. Because of this focused attention, and also because biological entities are inherently complex and dynamic, living beings will be "information-rich" sources for cognitive learning (Wilson, 1993). In addition, the biophilia hypothesis suggests that both positive and negative emotions—predisposing approach and avoidance, respectively—color perceptual/cognitive learning from living things and their environments. Consistent with this prediction is research showing that even infants are differentially attentive to and more positively engaged with living, as compared with toy, animals (Kidd & Kidd, 1987).

An evolutionary perspective indicates that certain small wild animals, such as spiders and snakes, as well as large predators would elicit children's fears, since such animals posed a threat in the environment of evolutionary adaptiveness. Studies of children's fears support this hypothesis (King, Hamilton, & Ollendick, 1998). In addition, research finds that environments that were associated with wild animal threat—darkness, deep woods, and being alone in unfamiliar places—also evoke children's fears (Heerwagen & Orians, 2002).

The biophilia hypothesis implies that children's attentiveness to wild animals would be heightened when the characteristics of living organisms—multisensory, dynamic, autonomous, coherent, and developing—are most evident. This would suggest that contexts that allow a full range of self-directed animal behavior would be more compelling than indirect, adult-mediated, and symbolic modes. Further, contexts that do not constrain the child's perceptions, emotions, and behaviors (except for child and animal safety) would be most enriching developmentally.

A second perspective derives from studies of the development of *folk-biology*, also called *naive biology*. This refers to "the cognitive processes by which people understand, classify, reason about and explain the world of plants and animals" (Coley, Solomon, & Shafto, 2002, p. 65). Such processes include intuitive understandings of living things, such as the construct of "alive," similarities between humans, (other) animals, and plants, and reasoning from a biological kind to biological attributes. Interestingly, no studies of folkbiology have explored how children (and adults) reason about and classify animals as wild versus nonwild—tame, domesticated, pets, and so on. Given that the notion of wild is a social construction embedded in a particular historical and cultural milieu, it would be interesting to determine how children come to acquire these classifications of animals and what they mean to the children themselves.

Susan Carey (1985) argues in an influential work that children's thinking undergoes a fundamental change from viewing humans as the prototypical animal, reasoning from humans to other animals, to viewing humans as one animal among others. (Carey's theory does not distinguish among species of animals.) Along with this shift is the gradual acquisition of concepts of *biological kinds*, what things are (or can be) alive, as well as *biological essentialism*, what mechanisms and processes underlie a biological kind.

While cognitive maturation plays a role in the development of biological reasoning, there is evidence that engagement with living animals prompts cognitive gains in biological concepts. Japanese kindergarteners who raised goldfish for a year were able to reason more accurately about the biological properties of the fish than were schoolmates without the goldfish care experience. Moreover, the children who had cared for goldfish also reasoned more accurately about unfamiliar animals, such as frogs, using analogies from their experiences with goldfish (Hatano & Inagaki, 1993; Inagaki, 1990). While their study is limited to pet care experiences, Alan Beck, Gail Melson, Patricia da Costa, & Ting Liu (2001) discovered that a ten-week home-based educational program for feeding wild birds at backyard bird feeders increased seven- to nine-year-olds' knowledge about wild birds.

While the evidence is limited, these results raise the possibility that observations of wild animals, perhaps combined with appropriate care experiences, may contribute to conceptual change in biological reasoning.

This possibility becomes more important against the backdrop of considerable ignorance and misinformation about animals found in surveys of children. For example, a survey of Connecticut youngsters (Kellert, 1997) found that 55 percent identified a whale as a fish, 79 percent believed that veal came from lamb, and a majority indicted animal predation as "wrong." Inner-city African American children had the least accurate information, while rural youngsters had the most. This might suggest that direct experience with animals (both wild and domestic) contributes to more accurate knowledge. It is likely, however, that children with the least knowledge also had access to fewer books, documentaries, and structured nature outings than did other children. Hence, it is unclear what modes of engagement with wild animals are linked to perceptual/cognitive changes. Would more distal and brief observation of living animals (say, at zoo exhibits) result in changes in biological reasoning similar to those of the Japanese schoolchildren who cared for goldfish for a year, in the Inagaki study cited above? Would mediated experiences of wild animals—watching a nature documentary or reading a book about monkeys—engage attention and be as information rich a source of learning as more direct, immediate engagement with living wild animals? The extent and kinds of engagement with animals that may impact biological reasoning are presently unclear.

Besides simple engagement with wild animals, attitudes of environmental stewardship, beliefs about similarities and differences between humans and animals, and the cultural valuing of nature may make animal behavior and characteristics more salient, and therefore focus attention for learning. Cultural values may predispose children to see the interconnections between or disjunctions among living things. Some research has shown that children who live in rural cultures emphasizing wilderness preservation and harmony with nature show advanced biological reasoning (compared to same-age rural children in other cultures). Thus, Menominee children in rural Wisconsin are more likely to accurately extend a biological property from animal to human than are their European American rural counterparts, who fail to do so on the grounds that "people are not animals" (Bang, Medin, & Atran, 2007, p. 13870). While the role of wild animals was not specifically studied, the investigators argued that Menominee children were reflecting their culture's emphasis on the close connection between people and wild animals. The Menominee creation story describes the people as coming from the bear, and Menominee society is organized into wild animal-based clans—bear, eagle, wolf, moose, and crane (Waxman, Medin, & Ross, 2007).

A third perspective derives from the theory of perceptual development offered by Eleanor Gibson (1988). In this view, children learn about the *affordances* of objects—the "what does this do?" of things—through observation and exploration. As Gene Myers (1998) demonstrated in his yearlong study of preschoolers and animals, the distinct affordances of various animal species, both wild and domestic, convey rich information about different ways of moving, sensing, responding, and acting. Marcelle Ricard and Louise Allard (1992) showed that even babies under one year of age react distinctively to an unfamiliar animal—in this case, a rabbit—as compared with an encounter with a novel toy or unfamiliar young adult. The infants crawled to approach and touch the rabbit, but looked and smiled at a woman only from a distance. They looked more frequently and longer at the rabbit than at a novel toy—a wooden turtle that moved, made noises, and flashed lights.

Different species communicate different affordances to children. In another study (Nielsen & Delude, 1989), researchers observed preschoolers and kindergarteners as they encountered the following live animals—a Mexican red-legged tarantula, an English Angora rabbit, a mature cockatiel, and a five-year-old female golden retriever—as well as two realistic stuffed animals—a dog and a bird. While 80 percent of the children never looked at the stuffed animals, the live animals attracted the children in distinctive ways. For example, 10 percent of the children touched the tarantula, which was on the floor in a terrarium, but 74 percent touched the dog, which was in a sit-stay position. Over two-thirds of the children talked to the bird, while only 16 percent talked to the rabbit, and none spoke to the tarantula. Thus, each species displayed its own distinctive way of being through its physical appearance and behaviors. The children, in turn, responded to these distinctive characteristics with distinct and adaptive ways of reacting to each animal.

As children observe a wide range of wild animal species, the distinct affordances of each species add to children's cognitive store of ways of being, thereby prompting children to respond in adaptive ways. In Piagetian terms, encounters with wild animals, whether immediate, mediated, or symbolic, all contribute to new schemata or cognitive constructs. All living animals present perceptually and cognitively rich, multisensory experiences, embodying novelty within recurring patterns of sight, sound, touch, smell, and movement. It is precisely this combination of new environmental information embedded within the familiar, what Jean Piaget (1960, p. 73) calls "moderate discrepancy from established cognitive schema," that most stimulates new conceptual thinking, or the emergence of new categories of thought, through the Piagetian process of *accommodation*. Furthermore, both Gibson and Piaget argue that child-directed, active exploration of the environment is the primary engine of perceptual-cognitive growth. It is through self-directed exploration that the child discovers the affordances of novel objects, the associations among stimuli—for instance, when I run toward the seagulls, they take flight—and the similarities and differences among animals. This suggests that spontaneous direct encounters with living wild animals result in the greatest developmental gains.

Self-development

The pervasiveness of animal symbols in every culture makes it evident that wild animals readily serve as symbols and expressions of aspects of the self. Psychodynamic theory, in particular, recognizes the power of animal imagery to reveal aspects of self. Sigmund Freud (1965) noted how frequently animals, including wild animals, appeared in the dreams of his adult patients. The centrality of animal imagery and associations is such that psychoanalytically oriented clinicians have devised projective tests focused on animal symbolization for both children and adults. One example is the Animal Attribution Story-Telling Technique (Arad, 2004), in which family members assign an animal counterpart to each person and then tell a story about the animal protagonists. According to Diana Arad (2004, p. 249), "the animal name attribution to family members . . . helps promote the description of personality traits and interpersonal relationships through the various animal counterparts." Animals also appear when children look at Rorschach inkblots. In one study of 650 children from two to ten years of age (Ames, Learned, Metraux, & Walker, 1952), half of all the inkblot interpretations involved animals, although not all were wild ones.

Children's readiness to use animal symbols to express and define aspects of the self is apparent in other areas. Detailed observations of young children's play uncover many instances of taking on animal identities. For example, a boy begins "growling" and stamping his feet while proclaiming: "I'm a gorilla!" to two other boys (Barker & Wright, 1951); preschoolers, after watching a robin, begin to "fly" and ask the teacher to catch them (Isaacs, 1948); after looking at a beetle, several boys play at being a beetle by crawling over each other while repeating "beetle, beetle, beetle" (Myers, 1998). Preschool children often dream about familiar wild animals, like frogs, birds, or deer, as well as farm animals, such as pigs, cows, and sheep (Foulkes, 1999). Wild animals, especially spiders, snakes, and sharks, appear on lists of top-ten fears reported by both rural and urban US children (White & Smith, 1989) as well as Australian children (Muris, Merckelbach, & Collaris, 1997) and adolescents (Lane & Gullone, 1999).

Animal symbols and imagery are so ubiquitous that psychologists have used them to assess important domains of children's human relationships. Thus, children's moral reasoning concerning *human* social relationships is usually measured by presenting moral dilemmas involving anthropomorphized animal characters, derived loosely from Aesop fables. In one scenario, for instance, a porcupine wants to spend the winter in the home of a family of moles. When the porcupine's quills begin to bother the tightly packed inhabitants of the moles' home, children in this study are asked, What should they do? (Johnston, 1988). The children's responses are interpreted as revealing their moral reasoning about a situation in which helping another *person* in need is pitted against self-interest.

As noted earlier, nowhere are wild animal symbols more pervasive than in children's media—books, videos, games, and the Internet. Animal characters allow children to imaginatively explore the boundary between human and animal, as Mowgli, the wolf boy in Rudyard Kipling's *Jungle Books*, does when he learns the ways of the animals' world, or as Stuart Little does when he inhabits the body of a mouse in E. B. White's children's novel of the same name. Issues of power and powerlessness play out for children as they read about small, wily wild animal "scamps and rascals"—Peter Rabbit, Curious George, or Cricket of *Cricket in Times Square*—whose antics get them into trouble, but who are all right in the end. As Paul Shepard (1996, p. 3) remarks, "Animals have a critical role in the shaping of personal identity." One might add that animals—wild and otherwise—reflect various facets of the child's sense of self.

Social Development
From an early age, children perceive animals—both pets and wild animals—as subjectivities, other autonomous living beings that, at least potentially, have intentions, feelings, and goals (Myers, 1998). It is not surprising, then, that when children encounter a wild animal, they react as if they are having a social encounter. In this way, wild animals, like other animals, are social beings for children.

Social psychological research on the development of social interactions suggests that qualities of contingency, indeterminacy, and intentionality in another are important (Fogel, 1993). A child therefore engages with another as a living social being when the other responds contingently to the child, in intelligible yet not completely predictable ways. Further, this other is seen as a social being when it appears to be autonomous, self-directed, intentional, and "coherent" (i.e., with an enduring self of its own despite variations in appearance and behavior). These qualities underlie children's (and adults') social responsiveness to robots, other technological emulations, and animals (Melson, Kahn, Beck, Friedman, Roberts et al., 2009; Melson, Kahn, Beck, & Friedman, 2009).

In pretend play, children often behave as if these qualities of social engagement exist by supplying them for the other. A stuffed animal thus can be made to "roar" and respond to the child's offer of a cup of tea with polite "drinking." Children also amplify existing capacities in the other for social responsiveness, as when a child pats a dog, articulates a verbal response for the dog—"Oh, so you want some more?"—and then pats the dog again.

The implications of this research are twofold. First, direct, unmediated encounters with wild animals, in which the animal responds contingently, are likely to be the most powerful in engaging with the animal as a social other. A child digging in the garden, for example, accidently unearths an earthworm that then wriggles away from their hand. Second, the power of pretense allows children to use this register to engage with wild animals *as if* they were playing a social role, even when the child is only observing the animal, and the animal is unaware of the child. Pretense also makes mediated and symbolic wild animal encounters potentially social ones as well—at least from the child's perspective. Yet the quality and intensity of a pretend, or "as if," social encounter is not equivalent to that of one in which both the self and the other respond in mutually contingent ways. Because the child controls both sides of a pretend social interaction, its power lies in the child's ability to rehearse and perfect interchanges. In an actual social interaction, the child must react to a dynamic interchange in which control is shared between the social partners. Thus, both direct and pretend social exchanges with wild animals have developmental utility, in different ways.

In the case of wild animals, as contrasted to pets or domesticated farm animals, direct social exchange with humans may sometimes be undesirable from an animal welfare perspective. Rather, observation of wild animals in a way that does not disturb them or their environment often is optimal. A child's encounter with a wild animal presents an opportunity for the child to recognize as well as respect the subjectivity and social capacities of the animal, while at the same time refrain from engaging those capacities in social interaction, when such interaction might adversely affect the animal's welfare.

Animals play a role in social development in another way, too. When children encounter an animal in the presence of other humans—adults or children—the animal frequently serves as a joint focus of attention and a catalyst, or what I have called "social glue" (Melson, 2001, p. 85) for interchanges among the humans. For example, in the study of wild bird feeding cited above (Beck et al., 2001), one child in each participating family was targeted for possible changes in bird knowledge and environmental attitudes. In the course of the ten-week program, though, involvement in wild bird feeding and "talking about the birds" increased for all family members, and persisted one year after the program's termination. In another study, my student Laura Richards observed a first-grade classroom with a variety of classroom "pets"—guinea pigs, gerbils, hamsters, a rabbit, a newt, a hedgehog, and fish—over a four-month period, noting interactions among the children focused on the animals. On average, during a single hour, twenty-one instances of social glue occurred (Melson, 2001). Although all the animals in this classroom were being kept as pets, it is likely that when children pay attention to wild animals jointly with other children or adults—for instance, observing zoo animals or noticing deer on a forest path—these encounters prompt discussion and social interaction among the humans.

Moral Development

Theory and research on children's moral reasoning and behavior have long centered exclusively on the morality of human–human relationships. This narrow focus has persisted until recently, despite the fact, as noted earlier, that children's moral development with respect to human relationships is routinely assessed with animal characters. Nevertheless, there is growing recognition that a morality of human–animal relationships and human–environment relations is developing as well. Across cultures, children are wrestling with issues of animal welfare and environmental degradation. Kahn (1999) interviewed African American youngsters in inner-city Houston and found them to be concerned with these issues.

A developmental progression in values concerning nature has been identified (Kellert, 1996), in which children gradually shift from predominantly egocentric attitudes—nature serving one's needs and desires—to recognizing the aesthetic, emotional, and symbolic values of nature (Kellert, 2002). In addition, experiences with nature have been shown to enhance moral reasoning in this domain. Direct experience with living animals is associated with less egocentric and dominionistic reasoning (Kellert, 2002), and more biocentric reasoning (Kahn, 1999). Moreover, individual differences in empathy and perspective taking predict attitudes toward environmental protection, including reasoning about wild animal welfare. When empathy and perspective taking increase (through role-taking instructions), for example, adults exhibit more positive attitudes toward protecting whales (Shelton & Rogers, 1981). Increased empathy, induced through such instructions, prompts more ecocentric moral reasoning (i.e., reasoning based on benefits to the environment) even when the object of reasoning is a vulture (Berenguer, 2010). As noted earlier, children from cultures emphasizing harmony with nature (and specifically, with wild animals) show more advanced biological reasoning. This suggests that children in such cultures also may reason about the moral claims of nature, wilderness, and wild animals in more biocentric or eco-centric ways, since both cognition and emotion undergird moral reasoning and behavior.

It is unclear, however, whether children's moral reasoning differs concerning specific *components* of nature—different types of animals, plants, streams, and mountains. For example, studies of folkbiology find that children recognize animals as "alive" and having a biology at earlier ages (around age four) than they recognize the same attributes in plants (around age eight), perhaps because plants do not engage in autonomous and apparently intentional movement—a hallmark of a living being for children (Coley et al., 2002). This suggests that if moral regard is linked to being viewed as alive, young children may be less concerned about the welfare or protection of plants than of animal life. With respect to different types and species of animals, mammals and those higher on the phylogenetic scale evoke more concern for animal welfare and preservation of wilderness habitats. Wild animals with what might be termed "privileged status"—panda bears, dolphins, porpoises, and whales—galvanize societal concern and protection in ways that equally endangered species, such as wolves and sharks, do not. Therefore, moral reasoning about wild animals appears to vary not only by the developmental level of the child, and the child's experiences with wild animals, but also by species, and by cultural views of wild animals and their habitats.

Conclusion

Any exploration of children's ideas and behaviors with respect to wild animals must be speculative as well as tentative. As noted above, in many areas, empirical studies are lacking, and in others, the few existing studies are of adults only. Cross-cultural investigations are few, despite clear evidence that engagement with wild animals is culturally mediated. Most assessments of attitudes toward animals fail to distinguish between types or species of animals, particularly companion versus wild animals. In general, sources of variation in children's contexts have not been investigated. Do children living in or near wilderness areas, for example, differ from other children in their cognitions, attitudes, or behavior related to wild animals, especially those living in these wilderness areas? What is the relation, if any, between a child's bond with their pet and ideas about or behaviors toward wild animals? Finally, variations in children's contacts with wild animals—immediacy, structure, frequency, duration, or context—need to be systematically linked to variations in child outcomes, such as biological knowledge, or attitudes toward wild animal welfare and conservation. While questions currently outpace answers when one focuses on children and wild animals, the developmental significance of asking those questions should be apparent.

Still, even considering the nascent state of our understanding of child-to-wild-animal relationships, one may advance some conclusions to guide the development of theory-driven developmental research. First, despite the widespread view that the wild has vanished from the world of childhood, contemporary children, even in industrialized, urban settings, have extensive and varied contacts with wild animals. Among the many worlds that children live in, they continue to live in the wild. Second, children's engagement with wild animals and the wild is rooted in human evolutionary history as well as cultural and historical forces. Wild animals and wilderness, for these reasons, should be presumed to be developmentally significant.

Engagement with wild animals is also multifaceted and diverse; it varies along dimensions of immediacy, structure, affect, and adult mediation. Then too, while children are increasingly experiencing wild animals indirectly, in mediated contexts designed by adults, these experiences nonetheless can be cognitively and emotionally intense. And despite the power of all forms of encounters with wild animals, both theory and research suggest that direct, unstructured engagement is most developmentally significant.

Lastly, there is accumulating evidence that engagement with animals, wild and nonwild, is linked to perceptual, cognitive, self, social, emotional, and moral development. Animals, wild and nonwild, focus attention, aid perceptual discrimination, foster accommodation (again, in Piagetian terms, the acquisition of new conceptual categories), reflect and refract the self, act as social others, and prompt moral reasoning about other species and one's place in the universe. Children's place is in the wild, and the wild is in children.

References

Ames, L. B., Learned, J., Metraux, R. W., & Walker, R. N. (1952). *Child Rorschach responses: Developmental trends from two to ten years.* New York, NY: Paul B. Hoeber.

Arad, D. (2004). If your mother were an animal, what animal would she be? Creating play-stories in family therapy: The Animal Attribution Story-Telling Technique (AASTT). *Family Process, 43,* 249–263.

Bang, M., Medin, D. L., & Atran, S. (2007). Cultural mosaics and mental models of nature. *Proceedings of the National Academy of Sciences of the United States of America, 104,* 13868–13874.

Barker, R. G., & Wright, H. F. (1951). *One boy's day: A specimen record of behavior.* New York, NY: Harper and Brothers.

Beck, A. M., Melson, G. F., da Costa, P. L., & Liu, T. (2001). The educational benefits of a ten-week home-based wild bird feeding program for children. *Anthrozoos, 14,* 19–28.

Berenguer, J. (2010). The effect of empathy in environmental moral reasoning. *Environment and Behavior, 42,* 110–134.

Berger, J. (1991). *About looking.* New York, NY: Vintage Books.

Boyd, N. A., & Mandler, G. (1955). Children's responses to human and animal stories and pictures. *Journal of Consulting Psychology, 19,* 367–371.

Bronfenbrenner, U. (1979). *The ecology of human development.* Cambridge, MA: Harvard University Press.

Carey, S. (1985). *Conceptual change in childhood.* Cambridge, MA: MIT Press.

Coley, J. D., Solomon, G.E.A., & Shafto, P. (2002). The development of folkbiology: A cognitive science perspective on children's understanding of the biological world. In P. H. Kahn, Jr., & S. R. Kellert (Eds.), *Children and nature: Psychological, sociocultural, and evolutionary investigations* (pp. 65–91). Cambridge, MA: MIT Press.

Decety, J., & Jackson, P. L. (2004). The functional architecture of human empathy. *Behavioral Cognitive Neuroscience, 3,* 71–100.

Drews, C. (2002). Attitudes, knowledge, and wild animals as pets in Costa Rica. *Anthrozoos, 15,* 119–136.

Dunlap, J., & Kellert, S. R. (1994). Zoos and zoological parks. In W. Reich (Ed.), *Encyclopedia of bioethics.* New York, NY: Macmillan.

Fine, A. H. (Ed.). (2006). *Handbook of animal-assisted therapy: Theoretical foundations and guidelines for practice* (2nd ed.). New York, NY: Academic Press.

Fogel, A. (1993). *Developing through relationships: Origins of communication, self, and culture*. Chicago, IL: University of Chicago Press.

Fogel, A., Greenspan, S., King, B. J., Lickliter, R., Reygadas, P., Shanker, S. G., Toren, C. (2008). A dynamic systems approach to the life sciences. In A. Fogel, B. J. King, & S. G. Shanker (Eds.), *Human development in the twenty-first century: Visionary ideas from systems scientists* (pp. 235–253). New York, NY: Cambridge University Press.

Foulkes, D. (1999). *Children's dreaming and the development of consciousness*. Cambridge, MA: Harvard University Press.

Freud, S. (1965). *The interpretation of dreams*. New York, NY: Avon.

Gibson, E. J. (1988). Exploratory behavior in the development of perceiving, acting, and the acquiring of knowledge. *Annual Review of Psychology, 39*, 1–41.

Goffman, E. (1963). *Behavior in public places: Notes on the social organization of gatherings*. New York, NY: Free Press.

Hatano, G., & Inagaki, K. (1993). Desituating cognition through the construction of conceptual knowledge. In P. Light & G. Butterworth (Eds.), *Context and cognition: Ways of learning and knowing* (pp. 115–133). Hillsdale, NJ: Lawrence Erlbaum Associates.

Heerwagen, J. H., & Orians, G. H. (2002). The ecological world of children. In P. H. Kahn, Jr., & S. R. Kellert (Eds.), *Children and nature: Psychological, sociocultural, and evolutionary investigations* (pp. 29–63). Cambridge, MA: MIT Press.

Inagaki, K. (1990). The effects of raising animals on children's biological knowledge. *British Journal of Developmental Psychology, 8*, 119–129.

Isaacs, S. (1948). *Intellectual growth in young children*. London, UK: Routledge and Kegan Paul.

Johnston, K. (1988). Adolescents' solutions to dilemmas in fables: Two moral orientations—two problem-solving strategies. In C. Gilligan, J. Ward, J. Taylor, & B. Bordige (Eds.), *Mapping the moral domain* (pp. 49–72). Cambridge, MA: Harvard University Press.

Kahn, P. H., Jr. (1999). *The human relationship with nature: Development and culture*. Cambridge, MA: MIT Press.

Kellert, S. R. (1996). *The value of life*. Washington, DC: Island Press.

Kellert, S. R. (1997). *From kinship to mastery: Biophilia in human evolution and development*. Washington, DC: Island Press.

Kellert, S. R. (2002). Experiencing nature: Affective, cognitive, and evaluative development in children. In P. H. Kahn, Jr., & S. R. Kellert (Eds.), *Children and nature: Psychological, sociocultural, and evolutionary investigations* (pp. 117–151). Cambridge, MA: MIT Press.

Kellert, S. R. (2009). Introduction to the special issue on children and nature. *Journal of Developmental Processes, 4*, 3–5.

Kellert, S. R., & Wilson, E. O. (Eds.). (1993). *The biophilia hypothesis*. Washington, DC: Island Press.

Kidd, A. H., & Kidd, R. M. (1987). Reactions of infants and toddlers to live and toy animals. *Psychological Reports, 61*, 455–464.

King, N. J., Hamilton, D. I., & Ollendick, T. H. (1998). *Children's phobias: A behavioral perspective.* New York, NY: Wiley.

Konner, M. (2005). Hunter-gatherer infancy and childhood: The! Kung and others. In B. S. Hewlett & M. E. Lamb (Eds.), *Hunter-gatherer childhoods: Evolutionary, developmental, and cultural perspectives* (pp. 19–64). New Brunswick, NJ: Transaction Publishers.

Lane, B., & Gullone, E. (1999). Common fears: A comparison of adolescents' self-generated and fear schedule survey generated fears. *Journal of Genetic Psychology, 160*, 51–58.

Louv, R. (2006). *Last child in the woods: Saving our children from nature-deficit disorder.* Chapel Hill, NC: Algonquin Books of Chapel Hill.

Lystad, M. H. (1980). *From Dr. Mather to Dr. Seuss: Two hundred years of American books for children.* Boston, MA: G. K. Hall.

Melson, G. F. (2001). *Why the wild things are: Animals in the lives of children.* Cambridge, MA: Harvard University Press.

Melson, G. F. (2007). Children in the living world: Why animals matter for children's development. In A. Fogel, B. J. King, & S. G. Shanker (Eds.), *Human development in the twenty-first century: Visionary ideas from systems scientists.* New York, NY: Cambridge University Press.

Melson, G. F., Kahn, P. H., Jr., Beck, A., & Friedman, B. (2009). Robotic pets in human lives: Implications for the human-animal bond and for human relationships with personified technologies. *Journal of Social Issues, 65*, 545–567.

Melson, G. F., Kahn, P. H., Jr., Beck, A., Friedman, B., Roberts, T., Garrett, E., & Brian, T. G. (2009). Children's behavior toward and understanding of robotic and living dogs. *Journal of Applied Developmental Psychology, 30*, 92–102.

Muris, P., Merckelbach, H., & Collaris, R. (1997). Common childhood fears and their origins. *Behaviour Research and Therapy, 35*, 929–937.

Myers, G. (1998). *Children and animals: Social development and our connections to other species.* Boulder, CO: Westview Press.

Myers, O. E., Jr., & Saunders, C. D. (2002). Animals as links toward developing caring relationships with the natural world. In P. H. Kahn, Jr., & S. R. Kellert (Eds.), *Children and nature: Psychological, sociocultural, and evolutionary investigations* (pp. 153–178). Cambridge, MA: MIT Press.

Nabhan, G. P., & Trimble, S. (1994). *The geography of childhood: Why children need wild places.* Boston, MA: Beacon Press.

Nelson, R. (1993). Searching for the lost arrow: Physical and spiritual ecology in the hunter's world. In S. R. Kellert & E. O. Wilson (Eds.), *The biophilia hypothesis* (pp. 201–228). Washington, DC: Island Press.

Nielsen, J. A., & Delude, L. A. (1989). Behavior of young children in the presence of different kinds of animals. *Anthrozoos, 3*, 119–129.

Oswald, L. J. (1995). Heroes and victims: The stereotyping of animal characters in children's realistic animal fiction. *Children's Literature in Education, 26,* 135–149.

Palmer, C. (2003). Placing animals in urban environmental ethics. *Journal of Social Philosophy, 34,* 64–78.

Phillips, C.J.C., & McCulloch, S. (2005). Student attitudes on animal sentience and use of animals in society. *Educational Research, 40,* 17–24.

Piaget, J. (1960). *The child's conception of the world.* London: Routledge.

Pyle, R. M. (1993). *The thunder tree: Lessons from an urban wildland.* Boston, MA: Houghton Mifflin.

Ricard, M., & Allard, L. (1992). The reaction of 9-to-10-month-old infants to an unfamiliar animal. *Journal of Genetic Psychology, 154,* 5–16.

Shapiro, K., & Copeland, M. W. (2005). Toward a critical theory of animal issues in fiction. *Society and Animals, 13,* 343–346.

Shelton, M. L., & Rogers, R. W. (1981). Fear-arousing and empathy-arousing appeals to help: The pathos of persuasion. *Journal of Applied Social Psychology, 11,* 366–378.

Shepard, P. (1995). *The others: How animals made us human.* Washington, DC: Island Press.

Shepard, P. (1996). *Traces of an omnivore.* Washington, DC: Island Press.

Sorge, C. (2008). The relationship between bonding with nonhuman animals and students' attitudes toward science. *Society and Animals, 16,* 171–184.

Thomas, E. M. (2006). *The old way: A story of the first people.* New York, NY: Farrar, Straus and Giroux.

Tunnicliffe, S. D., Gatt, S., Agius, C., & Pizzuto, S. A. (2008). Animals in the lives of young Maltese children. *Eurasia Journal of Mathematics, 4,* 215–221.

Turley, S. K. (2001). Children and the demand for recreational experiences: The case of zoos. *Leisure Studies, 20,* 1–18.

Varga, D. (2009). Babes in the woods: Wilderness aesthetics in children's stories and toys, 1830–1915. *Society and Animals, 17,* 187–205.

Vygotsky, L. (1978). *Mind in society.* Cambridge, MA: Harvard University Press.

Waxman, S., Medin, D. L., & Ross, N. (2007). Folkbiological reasoning from a cross-cultural developmental perspective: Early essentialist notions are shaped by cultural beliefs. *Developmental Psychology, 43,* 294–308.

White, P. N., & Smith, D. J. (1989). Content and intensity of fears in middle childhood among rural and urban boys and girls. *Journal of Genetic Psychology, 150,* 51–58.

Wilson, E. O. (1984). *Biophilia.* Cambridge, MA: Harvard University Press.

Wilson, E. O. (1993). Biophilia and the conservation ethic. In S. Kellert & E. O. Wilson (Eds.), *The biophilia hypothesis* (pp. 31–41). Washington, DC: Island Press.

6

Living Out of Our Minds

G. A. Bradshaw

Out of my experience . . . one fixed conclusion dogmatically emerges, and that is this, that we with our lives are like islands in the sea, or like trees in the forest. The maple and the pine may whisper to each other with their leaves, and Conanicut and Newport hear each other's fog horns. But the trees also commingle their roots in the darkness underground, and the islands also hang together through the ocean's bottom. Just so there is a continuum of cosmic consciousness, against which our individuality builds but accidental fences, and into which our several minds plunge as into a mother-sea or reservoir. Our "normal" consciousness is circumscribed for adaptation to our external earthly environment, but the fence is weak in spots, and fitful influences from beyond leak in, showing the otherwise unverifiable common connexion.

—William James, *Essays in Psychical Research*

Physical fences were no accident in James's world. Miners, trappers, and settlers poured across the continent, ax blows silenced murmuring pines and hemlocks, and miles of fences converted seamless landscapes into complex quilts of ownership. By the time the psychologist died on the eve of World War I, the US wilderness was well under way to domestication.

My awakening to James's cosmic consciousness occurred in another land, far away, in Africa. The colonial grip there has also tried to twist nature into submission. Damage is pervasive. In the South and East, the most noticeable impact is on wildlife. Postcontact numbers are but a tiny fraction of what they used to be, and even the iconic elephant teeters on the edge of extinction. Yet civilization seems to have a harder time retaining its mark. There is something formidable in Africa's nature that defies complete surrender. Trucks fail to withstand penetrating red dust and ruts, and try as they might, Western fashion styles lose their identity of origin. Baseball caps and T-shirts reduce to mere accents when adorned. Africa is powerful magic.

The flicker of revelation ignited one afternoon, while staying at a South African resort. We sat in desultory conversation after the gusto of lunch. Day hung in the golden hours between morning's vigor and the wary vigil of night—lions slept, baboons groomed, and wildebeest tails switched idly. The low laughter of maids and gardeners carried across spacious lawns. I rose and left to fetch a notebook, but instead, distracted, wandered up to the gift shop.

A young woman sat with legs crossed on a stool at the lintel. She wore a scarf, turban style and colored pale yellow, matching her dress. Her eyes flashed fresh white against dark brown. We exchanged smiles, and she gestured for me to come in. The shop was filled with the usual carved elephant statues, lion- and hippo-emblazoned T-shirts, postcards, maps, and trinkets for visitors hungry to possess a piece of Africa. I spied a wooden rhino standing defiantly on a shelf. Head turned slightly toward the viewer, his armored body was painted in the bright lacquered colors of South Africa's flag: red, green, blue, gold, and black. I bought the rhino and paid in fumbling currency, making an offhand remark about the beautiful countryside. The woman asked where I was from, and within minutes we sat beside each other, laughing in touch and voice. I have no recollection what we talked about, but she pressed me to come by the kitchen for tea later that day. Her name was Ngina.

By now my companions had emptied their glasses and stood outside the waiting truck. Within a few minutes, we were off to look for lions. We were part of a group of scientists looking into the effects of a large-scale wildlife reintroduction. After extirpation in past decades, lions and the other "big-five" wildlife were being imported into the country to stock parks for ecotourism. Postapartheid South Africa was readying to regain its former status as a popular tourist destination.

The truck made its way along brush-lined dirt roads, bumping along until arriving at a knobby kopje, a small hill of jumbled boulders and prickly bush. We pulled over, jumped out, and headed up the steep incline. Finally making it to the top, the veterinarian in the group handed me binoculars and pointed out over the plain to a pair of eagles circling above. There they were, looking larger, darker, and more entitled than their North American cousins. I breathed in the open blue sky in awe of Africa's wilderness.

It wasn't the time or place for lions, so we returned for dinner. After reaching my room, I threw down my pack and dug through the suitcase for clean clothes. I washed off the dust, pulled on a fresh shirt, and hurried down to the kitchen. Ngina ran up with a big smile and clasped my forearms, pulling me into the *rondoval*. Banging pots and clattering ladles blended with music blasting from a small radio on a window shelf. Faces looked up and hands waved as Ngina introduced me to the kitchen staff, then led me to a small table in the corner, where we sat for tea.

She told me about her family, her father who had gone to work in the mines, and her mother who had stayed with the four children, waiting for her husband's return while doing laundry sent down from the big house. Ngina then told me about her brother who was killed in the streets. "Times were hard," she said. The conversation stopped, and the kitchen sounds dinned on. Ngina stirred her tea and offered me more. I asked, "How did you manage?" Carefully putting down her cup, she fell silent, and then dipping her head to catch my gaze, she held my hands and replied, "It was difficult. But I chose life over survival."

Time suspended, and for a moment, fences dropped. Suddenly, the bark of laughter from arriving guests cut through the air, and we materialized back into separate shells of reality. Someone called, and Ngina slipped from serious to gracious warmth. Excusing herself, she got up, gave me a strong embrace, and looking back with a smile, went off to help serve cocktails. I joined colleagues under the sky awash with stars.

The next day we were gone. I had looked for Ngina and asked of her whereabouts to say good-bye, but the kitchen staff said that it was her day off. I never saw her again. Her story stayed with me. Every so often, I would think back to what she had said about her choice of "life over survival." It wasn't until years later that I felt that I had understood. Unsurprisingly, insight came out of Africa, again, from the stories of other, displaced compatriots.

Fencing the Wild

Alterations of personal identity were a constant feature of the [concentration camp] survivor syndrome. While the majority of . . . patients complained "I am now a different person," the most severely harmed stated simply "I am not a person."
—Judith Herman, *Trauma and Recovery*

Over the past decade, my work has focused on wild victims (Bradshaw, 2009). There was Echo, the African elephant matriarch, striding across Kenyan plains with her family as they struggle to navigate landscapes made lethal and perforated by an ever-consuming human population—and its guns—made greedy in the race for ivory. There were also the young elephant bulls whose serial killings of over a hundred rhinoceroses broke through the barrier of anecdote to scientific truth of an animal society gone mad from human violence (Bradshaw, 2005, 2009). And then there were the captives, wrenched away and held behind barriers to stem their wildness. Wildlife captivity is so deeply ingrained in our culture that we think it nothing to parade a multiton herbivore that could obliterate all those around them in literally the blink of an eye. The normal restraint of elephants derives from a mixture of a pacific nature and spiritual obliteration. Former zookeeper Ray Ryan (quoted in Bradshaw, 2009, p. 213) observes:

Circus elephants—walking in a single file down a stress. Why don't they run away? It's simple. It's because they're dead. They are dead souls in circuses and zoos. The only way you get elephants who are so powerful to do what a human wants with just a flick of their hand, is to beat the soul out of them. I saw it in Peach [an elephant at the San Diego Wild Animal Park], when I beat her, I saw her soul leave.

Stories of captive elephants begin with violence, and almost all end in violence. Black Diamond was an awesome bull elephant who, but for capture, would have become a leader and grandfather of a proud lineage many times over. Captive elephants are brutalized daily with beatings, electric shocks, and chaining, and Black Diamond's fate was no different. One day, however, nature would not withstand the human mind.

In fall 1929, in Corsicana, Texas, Black Diamond met up with a former trainer, H. D. "Curley" Pritchett (referred to elsewhere as Curley Pickett), who had brought along a lady friend to show off his erstwhile charge. According to accounts, Black Diamond suddenly wrapped his trunk around Pritchett and threw the man over a train car, then trampled and killed the woman. John Ringling, the elephant's "owner," promptly "condemned Black Diamond to death"—with no trial and certainly no jury by peers. People first tried to poison the canny bull with cyanide-laced peanuts and oranges. When this did not succeed, three men stood facing the bull, who stood tethered and weighted with steel chains. Black Diamond "howled, tried to jerk away . . . wildly trumpeted and twisted," while over 60 (some accounts say more than 170) bullets were fired before he buckled to his knees and died (Animals, 1929). The elephant's head was skinned and mounted for display. Years later, a local businessman had Black Diamond's remains dug up to obtain the skull and now exhibits it with the bull's mounted head (RoadsideAmerica.com, n.d.).

Medundamelli (also known as Dunda) provides another glimpse into life behind fences. At age two, the baby Dunda was captured after her family was slaughtered in Africa. She arrived at the San Diego Wild Animal Park and Zoo on September 18, 1971. In 1989, the head zookeeper for elephants and four other men sought to "discipline" the female elephant. While all four of her legs were chained to the extent that she was nearly prone, the men beat her with wooden ax handles. The authorities considered this behavior an accepted technique: "You have to motivate them and the way you do that is by beating the hell out of them" (quoted in Bradshaw, 2009, p. 189). Only after the second day, when Dunda accepted an apple, did her punishment cease. A senior trainer shared what he heard from one of the perpetrators:

I asked how badly Dunda was hurt, and [he] replied, "Well, she rolled over on her side and moaned." I was appalled when I heard this and I still can't get it out of my mind. Lou described the area around Dunda as covered with the horrific flow of liquid feces. This condition of excessive elimination is a sure sign that the animal was in a state of fear and is easily recognized. An animal being disciplined for aggression would not do this until it had been frightened by the disciplinary action and would understand what was happening. Dunda was defecating and urinating profusely from the beginning. . . . This frightened, cornered animal was in a state of panic situation and had no idea what was happening to her. She was fighting for her life, and she must have felt it was over. (ibid., p. 193)

Dunda's wound showed a "deep crevasse in the left forehead . . . so deep that it reminds you of the time Alan Roocroft [the head elephant trainer and zookeeper participating] described the use of sledge hammers to discipline elephants in Hamburg" (ibid.). Bruce Upchurch (ibid., p. 194), an elephant keeper visiting from the Woodland Park Zoo in Seattle, Washington, said that he did not hit Dunda, but that he "did not see anything that shocked or upset him," and "when necessary, the Seattle Zoo uses the same type of discipline." A few years later, Dunda was moved to the Oakland Zoo, where she remains today still behind fences.

Ray Ryan (ibid., pp. 211–212) did not participate in Dunda's beating but speaks of humans' tempest within:

It's hard to describe, but when you eventually get control over someone who has no natural control and is so big, well, it makes you feel big. It is a real display of machismo. . . . You could show you were a real man if you could beat down a big powerful animal. And I could always tell who had had a fight with their wife the morning or night before. We have not changed much since cave days. Men are still beating up women, still trying to run the world with domination. And if you notice, all the elephants we work with are females.

This theme of gender, sexuality, and violence is repeated behind other fences.

After arriving under the influence of alcohol, an elephant keeper at the Oregon Zoo chased and abused Rose-Tu, a five- to six-year-old female child elephant. According to testimony, the keeper swore at her, and stuck an ankus (a metal pronged tool used to punish and control elephants) into her anus and pulled hard, after which she crumpled to the ground. He was eventually stopped, but not before Rose-Tu was wounded more than 170 times. Physical scars trace those inside. After being mated with a bull elephant and giving birth, Rose-Tu trampled her infant. Infanticide is not uncommon in captivity. Other practices banned by the Geneva Convention are routine in zoos (Capaldo & Bradshaw, 2011).

Nonconsensual captive breeding is increasingly common in zoos because of the drive to replenish "stock." On average, elephants in captivity live half as long as their free-living counterparts (Clubb, Rowcliffe, Lee, Mar, Moss, & Mason, 2008). Wildlife progeny are born into concrete jungles through masturbation by humans or other methods that "train" orcas and elephants to yield semen for artificial insemination (YouTube, 2011a). Holding up a plastic bag while speaking to the camera, an orca trainer at SeaWorld describes the process: "When he [the orca] sees this bag, he'll roll over and present his penis and ideally he'll emit that semen sample" (YouTube, 2011b).

Chai, an Asian elephant at the Woodland Park Zoo in Seattle, has been inseminated fifty-seven times. "Experts" from Germany were brought in to perform the procedure in 2005. To execute the insemination:

A scaffold was erected at the business end of Chai so Dr. Thomas Hildebrandt, from the Berlin Institute for Zoo Biology and Wildlife Research, had easy access. Hildebrandt was covered in plastic protective gear and sporting a bike helmet he had equipped with ultrasound imaging goggles. Beneath Hildebrandt, his colleague Dr. Frank Goeritz sat on a stool in front of a bank of computer screens, electronic equipment and a tangle of computer and power cords. Tended by six or seven keepers who were there to keep both the Germans and Chai happy, the procedure involved Hildebrandt inserting an ultrasound probe into the elephant's rectum while Goeritz fed an endoscope (a light-emitting tube that allows visual inspection internally) into a larger catheter that had been inserted into Chai's "vestibule." The vestibule is just one feature of the animal's 10-foot-long reproductive tract that makes artificially inseminating this creature a challenge. An elephant's vaginal opening is not external but located inside this chamber called the vestibule. The vaginal opening is about dime-sized with two false openings, or pockets, on either side. And the bladder's opening is also nearby, and much larger, within the vestibule. (Paulson, 2005)

The zoo's general curator, Dr. Nancy Hawkes (quoted in Phinney-Wood.com, 2011), considers that "artificially inseminating an elephant is a technique that enhances animal welfare." Bruce Bohmke (quoted in Klein, 2011), zoo chief operations officer, states that forty-eight of the total were

inseminations [that] were done more than 10 years ago. They weren't real inseminations. They were attempts to just inseminate by putting semen into the elephant, which was not, as we know now through studies in captivity, an effective way to inseminate them. . . . We have an effective way to inseminate them now and we have had at least one pregnancy because of that, which unfortunately ended up in a miscarriage.

According to Hawkes (quoted in PhinneyWood.com, 2011), the last two decades of failed inseminations occurred because "they were usually just inseminating the bladder." One can be sure that Chai knew the difference.

Personalized violence against captive wildlife extends to all species. Jeannie was also captured from the African wilderness. She was a beautiful chimpanzee, *Pan troglodytes*, who endured decades of brutalization as a biomedical subject. Jeannie was born in 1975. Before rescue and arrival at the Fauna sanctuary in Quebec, Canada, where she lived out her final days, Jeannie was kept at the New York Laboratory for Experimental Medicine and Surgery in Primates (LEMSIP). Seven years before, in 1981, she had been donated by Merck Sharpe and Dohme (Merck) pharmaceuticals to the Buckshire Corporation. For nine years, she sustained invasive research, including repeated vaginal washes; multiple cervical, liver punch, wedge, and lymph node biopsies; and infection with HIV, hepatitis NANB, and C virus (Bradshaw, Capaldo, Lindner, & Grow 2008, 2009). She was also used in rhinovirus vaccine studies and experienced over two hundred "knockdowns" (i.e., anesthetization by a dart gun). The LEMSIP staff described how Jeannie suffered "a nervous breakdown" and multiple emotional issues. Psychotropic medications were used to quell these symptoms, which included self-injury, seizurelike episodes, screaming, and alternating trancelike and highly anxious states.

Gloria Grow, director of Fauna, visited Jeannie at LEMSIP prior to her rescue. She explained how Jeannie began screaming and spinning in her cage when she caught sight of personnel approaching in masks and protective clothing. Her screams caused a chain reaction of terror among the other chimpanzees also held alone in 5' × 5' × 7' steel cages suspended along a fixed ceiling track. Jeannie would and could not stop screaming. She frothed, salivated, rolled her eyes back, urinated, and defecated, while striking her body against the sides of the cage.

Jeannie's condition was so severe that she was removed from studies and the staff considered her a candidate for euthanasia. The sanctuary intervened and convinced the lab to release Jeannie. At age twenty-two, she was sent to the Fauna sanctuary until her death on January 1, 2007. Jeannie retained a ritual from the lab while she lived in the sanctuary. She would often take her food and arrange each piece deliberately in a circle around her, creating an "inner sanctuary" of safety (Bradshaw et al., 2008). Her story is horrifyingly typical of other chimpanzees—hundreds of whom are still "in use" by laboratories in the United States (Capaldo and Bradshaw, 2011). Similar to concentration camp survivors, Jeannie and other chimpanzees used for research have been diagnosed with Complex Post-Traumatic Stress Disorder (Bradshaw et al., 2008, 2009).

The Space in Between
Nature is one, and to me the greatest delight of observation and study is to discover new unities in this all-embracing and eternal harmony.
—John Muir, quoted in Dreier, 1913

Stories of nature's wild children reveal wilderness as *someone*—beings other than stage props of an anonymous backdrop to human dramas. Nature has a face. Nature is ensouled. Such stories present living and dying testimony that cosmic consciousness still pulses. *God is alive; magic is afoot* (Cohen, 2001). Their power to move and live beyond the body speaks of a different consciousness—one that exists beyond survival.

While some cry anthropomorphic trespass when Chai's or Jeannie's experiences are referred to as rape, Black Diamond's as slavery and torture, and mass killings of elephants comparable to the European Holocaust (King, 2009), truth cannot be denied. Science itself has dismantled accidental fences by proclaiming human–animal comparability in psyche (Bradshaw & Sapolsky, 2006; Marino, 2011). Neuroscience confirms what Charles Darwin perceived over 150 years ago from his ethological studies. All vertebrates, humans included, share common structures and processes that govern emotion, cognition, sense of self, and consciousness. Calibrating human identity relative to other animals doesn't seem to make much sense anymore; the similarities far outweigh the differences (except in the realm of violence, where humanity outclasses every other species). Instead, the study of human and other animal minds and souls has merged with the new field of trans-species psychology (Bradshaw, 2005, 2009). Fences separating species are, like past models of consciousness, accidental.

At the same time, Jamesian consciousness has re-emerged in the language of neurons (Noë, 2009). According to philosopher-cum-neuroscientist

Alva Noë, science made a wrong turn when it claimed that the machinery inside our heads produced consciousness—felt existence and experience. Rather, insists Noë, what we thought was inside, is outside. Consciousness is not a tape playing inside but instead is created out of our minds with others; it is not "something that happens in us. It is something we do" (Noe 2008). Noë likens consciousness to a dancer in dialogue with their environment. What we experience and what we know forms in tandem with the matrix of relationships in which we are immersed. We are not only what we eat, but also who we see, touch, smell, taste, and hear. *The life is not inside the animal. The life is the way the animal is in the world* (Noë, 2008).

The theory of biological consciousness does not stand in isolation. It is one among a number of other theories that when taken together, provide conceptual mortar and empirical bricks to Noë's vision. Overall, science's theoretical lens has changed. Instead of reductionism's divvying up the world into smaller and smaller pieces like so many *matryoshkas*, current theories reverse this conceptual meiosis and reconcile differences. Nature is no longer pitted against nurture in the battle to determine who we become. Genes and the environment converse as we age and traverse the relational map of the world. Psychological development depends not only on what we inherit but the people and things we experience as well. Each of us is still unique, yet we take a little of the others whom we know with us—the good and the bad. Early relationships are thought to be most influential (Schore, 2005). If a child experiences abuse or premature separation from their parent, such as the rhinoceros-killing and copulating young elephant teens, they become traumatized, emotionally insecure, and psychologically vulnerable. Our brains, minds, and sense of selves mirror relationships. The self is relational. James's fences—the perception and belief that each of us stands separate from the natural matrix of life—are revealed as accidental, illusory, and ephemeral.

What we live is made up of moments, such as those shared with Ngina, collected and made meaningful over time in partnership. The cavalcade of relational moments gathered over a lifetime—a snippet of conversation, the flash of an image when eyes meet eyes, or the sensation when scent meets sensor—come together like so many water lilies to create a single coherent narrative. It is this experience that we call consciousness. Life does not happen in neat little organismic packets. Rather, life is the intake of breath of a shared world, and it happens out of our heads somewhere in between.

Life beyond Survival

The last of the human freedoms—to choose one's attitude in any given set of circumstances, to choose one's own way.
—Viktor Frankl, *Man's Search for Meaning*

This alternative understanding of consciousness puts the stories of Ngina and the others in a completely different light. Broken in body, scarred in mind, Jeannie, Ngina, and Dunda emerge not as survivors but instead as icons of defiant connection, each with a sense of self who experience life as something other than in the singular. Collective, interdependent cultures present a contrasting psychology to that of the West (Greenfield, Keller, Fugligni, & Maynard, 2003). Traditional tribal humans who doggedly refuse to embrace "civilization" along with elephants, orcas, birds, reptiles, insects, and others indifferently labeled as "biodiversity" envision their world quite differently. The sense of self has potential to be trans-species, extending beyond form. It is not anthropocentric but rather ecocentric, inclusive of nonhumans (Kirmayer, 2007). This understanding concurs with attachment theory: culturally and neurobiologically, our sense of self reflects the environment in which our minds develop, as in the case of cross-fostered (reared by someone other than their own species) chimpanzees, parrots, and humans (Candland, 1995; Bradshaw et al., 2009). Their reality does not seek to sever the thread of life. I is we, and species differences fade into mere cultural nuances (Savage-Rumbaugh, Wamba, Wamba, & Wamba, 2007; Bradshaw, 2010).

There is a choice between survival and living. Take Minnie. For four decades, she was chained and abused in a circus. Over time she took on the identity of her human trainers, and mirroring them, inflicted the pain that she endured on her elephant sisters. Her world of interconnection succumbed to the brutality of human survival, but finally, after being sent to a sanctuary, she was able and chose to try to rekindle her elephant self. The external containment provided in the sanctuary through psychological and physical safety supported internal containment that allowed expression of her self of origin—a prosocial collectively centered elephant self. Minnie in recovery was someone capable of love and devotion, as she exhibited in the tender grieving for her dead companion, Queenie. Carol Buckley, cofounder of the Elephant Sanctuary in Tennessee, describes Minnie as she "painstakingly and gently lifted one front foot and then the other over Queenie's bloated belly, placing both feet centimeters from Queenie's back. With grace and care, Minnie arranged her hugeness protectively and completely over her friend's body" (Bradshaw, 2009, p. 161). Minnie chose life and extricated herself from the internalized identity of (human) abuser acquired in the brutal past.

Similarly, Marco, a Moluccan cockatoo who lived in a two-foot-square cage in a windowless dark garage for three years, chose life. When released to a sanctuary, Marco refused to make any sound other than that of a closing door. Yet despite years of black loneliness, he lived beyond his pain and tenderly cared for a crippled conure, another refugee in the sanctuary. Marco, too, not only survived, he lived.

Primo Levi's moving tribute to an Auschwitz camp worker, Lorenzo, provides yet another example. Lorenzo was a mason employed in France when his firm along with its employees were swept up by the Germans in 1939 and transferred to Upper Silesia. Levi was selected by his kapo to assist Lorenzo and another man to rebuild mortar and brick walls that had been damaged by bombs. At first the two men did not recognize that they were both Italian, and even neighbors from Piedmont—one from Turin, and the other from Fossano. Despite the possible lethal consequences, they spoke together. Levi (1995, p. 123) describes how his life was saved with hidden soup and slices of bread, and how he and his companion were astounded that in "the violent and degraded environment of Auschwitz, a man helping other men out of pure altruism was incomprehensible, alien, like a savior who's come from heaven."

Levi tried to compensate Lorenzo for his help, but the man refused. "I offered to have some money sent to his sister, who lived in Italy, in exchange for what he did for us, but he refused to give us her address." He later learned that Lorenzo had helped several others, yet did not mention it for fear of seeming to brag. Lorenzo was indeed a savior, but a "morose savior, with whom it was difficult to communicate" (ibid., p. 113).

When the war ended, Lorenzo made it out and walked for months to Turin, where Levi's family lived. He lied to Levi's mother, saying that all the prisoners at Auschwitz were dead, thinking that in all likelihood this was the case and that it would be better for her to resign herself to her son's death. Levi's mother offered Lorenzo money so that he could take the train, but again he refused.

After his release from Auschwitz and then five months in Russia, Levi returned home to Italy. He went to Fossano, where he found Lorenzo living a nomadic life as a mason and sleeping outdoors even in the bitterest cold of winter 1945–46. "I found a tired man; not tired from the walk, mortally tired, a weariness without remedy." The writer had secured a masonry job in Turin for his comrade, but Lorenzo would not take it.

Soon thereafter, Lorenzo became ill, and with assistance from Levi's physician friends, was hospitalized. Yet Lorenzo ran away and died a few days later. Levi closes his story with these words: "He was assured and coherent in his rejection of life. He, who was not a survivor, had died of the survivors' disease" (ibid., p. 118).

Levi makes the point that the man he encountered after the war was "mortally tired," someone who succumbed to "survivor's guilt," symptomatic of many who managed to endure while their loved ones and others did not. Later in his book of essays on the camps, Levi writes that "the worst survived, that is, the fittest; the best all died" (quoted in Freilander & Landau, 1999, p. 564). Himself beset with depression and memories, Levi's death forty years later is officially considered a suicide. Despite the despair that possessed them both, the two men ultimately chose to live through the camps, not just to survive: "Lorenzo was a man; his humanity was pure and unblemished, and he was outside this world of denial. Thanks to Lorenzo I happened not to forget I myself was a man" (quoted in Judt, 1999). It may be that past horrors and what they revealed about present-day humanity became too much to bear, and overwhelmed Lorenzo and Levi—that life was reduced to survival.

These stories are not chronicles of survival eked out from misery but instead a fierce embrace of life. *Life*. Life is the common thread linking Ngina and Noë's neurophilosophical thesis. Ngina, Minnie, and the others who were forced to relate to life through experiencing near death all illustrate the difference between hard survival and pulsing life. They lived because they dared to live for someone or something outside themselves. Meaning sprang from the space in between. They chose life by finding meaning in something outside their own bodies, to live a relational existence that does not define itself within the confines of skin-covered existence.

From a Paradigm of Fear to a Paradigm of Life

What am I to do, Mama, when I am living a sane, but specially unincorporated life in so limited a society?
—S. Bogus, "Mom de plume"

From this perspective, biological consciousness has brought down the entire house of René Descartes. Together, dissolutions of species and mental barriers have ushered in a new ethical consciousness as well as theory of life. This understanding comes not a second too soon. Denizens of the modern–postmodern era are prisoners of survival living in a world of ones where motive and meaning telescope into the single social cell of the individual. Humans live as islands in a sea of fear, transfixed by the struggle to withstand what humanity itself has created and memorialized in the litany of genocides, wars, and ecological destruction (Narvaez & Gleason, 2012). Modernity is the recipient of that mixed blessing, "May you live in interesting times." Interesting indeed, if the definition includes environmental collapse, nuclear war, and psychological estrangement so profound as to make Post-Traumatic Stress Disorder epidemic. Even with the reassuring presence of thermostats that buffer climatic stress, refrigerators that smooth out erratic supplies of food, and other conveniences that earmark modernity, life feels fraught with uncertainty.

Impending disaster is nothing new. A glance back into human history shows that life has always been uncertain. There has always been that testy tribe next door itching to pillage, or a hot-breathed *Tyrannosaurus rex* flashing menacing teeth in anticipation of their next meal. Then there were hundred-year floods, locust hoards, and fluke accidents that never made it possible to completely relax. But today's uncertainty has a different cast than the past. There is not only a difference in scale— obliteration of the planet versus a village; antidotes to uncertainty—the familiar symbols and memories that used to bring some sense of comfort as well as stability in the wake of terror—also have themselves become threats.

Family intimacies and a schoolteacher's loving embrace are corroded by suspicion of sexual abuse. Blessings of Thanksgiving harvests are symbols masking the violent appropriation of indigenous land, and the recapitulation of past human genocides is diverted to millions of turkeys. Wine, timeless juice of the earth, seemingly unburdened by worries that beset other foodstuffs, has vanquished the cork that has graced Mediterranean soils even before the ancient Greeks. Almost every aspect of modern living seems to be fraught with contradictions.

There does not appear to be a way to escape this existential quandary. Even while standing at the grocery checkout, we are faced with deciding the planet's future: "Plastic or paper?" Using plastic for carry-home groceries saves dwindling forests and staves off greenhouse gases, but adds to the vast waste equivalent to the size of Rhode Island that has accumulated in the middle of the Pacific Ocean. No matter where we turn, what right we seek to amend, there seems to be a dark side that counters what good was intended; reality distorts in this cultural house of mirrors, and we are strangers in our own home.

Everyday inconsistencies and paradoxes overwhelm. The mind jams, it "does not compute," and in desperation resorts to artful evasions—cognitive dissonance, doubling, dissociation, sociopathy, and more common guises of alcohol and other addictions—to grasp some semblance of balance and meaning. When decisions devolve into Sophie's choice and familiar solutions cease to resolve problems, then we have the signs of a paradigm in crisis. External guideposts—the ideas, beliefs, cultural practices, and pundits that have provided the psychological, political, and practical architecture for knowing how to live—no longer align with what we experience or know. Which is why a new paradigm, a new way of seeing, understanding, and living in the world, is so desperately needed (Bradshaw, 2009). The choice to live or survive is the choice between two paradigms: "One can discern two trajectories in current history: one aiming toward hegemony, acting *rationally* within a lunatic doctrinal framework as it threatens survival; the other dedicated to the belief that 'another world is possible'" (Chomsky, 2003, p. 45). Biological consciousness offers this other world—an idea organically grown in the same minds and culture that created the problems to begin with. But an idea resonant with other philosophies that flourished when the earth was not darkened with the shadowy threat of nuclear war, when animal kin did not flee at the sound of a human voice.

A new theory of mind is not an attempt to render matters of conscience and ethics into equations and numbers; far from it. Further, it would be folly to regard science as a knight in shining armor. As psychological seer C.G. Jung (1964, p. 95) observed a half century ago, science cuts both ways—offering ingenuity, on the one hand, and a Faustian price, on the other:

As scientific understanding has grown, so our world has become dehumanized. Man feels himself isolated in the cosmos, because he is no longer involved in nature, and has lost his emotional "unconscious identity" with natural phenomena. . . . Thunder is no longer the voice of an angry god, nor is lightning his avenging missile. . . . No voice now speaks to man from stones, plants and animals, nor does he speak to them believing they can hear.

Noë's theory of biological consciousness brings welcome cheer that stones, snakes, and thunder all have not stopped talking; we have stopped listening. While humanity may not wish to hear, nature is speaking loud and clear in the language of drowning polar bears and the anger of unseasonable weather. Science's new understanding of consciousness and psyche is significant because it offers a way to connect what we knew to what we will discover. The ability to articulate the relationship between something as intangible as a worldview and the hands-on operations of thinking and method in customary symbols provides a reassuring cognitive trail of bread crumbs leading from the known world to an unfamiliar land unencumbered by past assumptions.

The wild ones emerge radiant, teaching the need for emulating, not pitying, those who have suffered so mightily. Their testimonies eschew monuments commemorating survival in the bloody past. Instead, they urge those of us who remain to live, not just survive. We are not alone. In the magical world of *The Golden Compass*, every person has a life companion in the form of an animal called a "daemon." A daemon is a cross between the soul and a best friend. The animal form it takes is a manifestation of the host's true personality (dopetype, 2008). *The Golden Compass* by Phillip Pullman (2003) is the story of a young girl, Lyra, undertaking a journey with her daemon, Pan (short for Pantalaimon, which is also Greek for all compassionate), in a mysterious and dangerous parallel universe. In this strange but familiar world, everyone is born with a daemon. A soul-friend may take multiple forms. Pan morphs from moment to moment from delicate moth to mighty eagle to tiny mouse. But all is not well for girl and daemon. There are dark forces at work that threaten. Lyra discovers that members of the General Oblation Board are conducting nefarious experiments that entail severing the daemon of kidnapped children. Violently separated from their daemons, the children as well as their allies wither and die. Along with her daemon and others, including a giant polar bear king, Lyra rushes through diverse adventures to save her friends and allies.

Pullman lyrically paints the bridge between the new, relational science, and intentionally or not, identifies the source of postmodern destruction in this beautiful yet haunting tale. The present-day, near-epidemic trauma gripping human and animals across continents has grown from acts based on human–nature separation. Trauma emerges not only through the rampant spread of asocial, amoral behavior and violence experienced at the hands of others but also as a self-inflicted embrace of Western thought and action (Narvaez & Gleason, 2012; Bradshaw, 2012a, 2012b). As surely as the blade sliced through the gossamer connection of child and daemon, the concepts and agenda defining Western society cuts humanity from its life source, its daemon, nature. The daemons that populate the natural world, those who make the untrammeled forests and mountains tingle with life and mystery, wither and die. The world has been fragmented into brittle edges and wounded psyches by humanity's contraction into the bare bones of survival. Images of drowning polar bears, tribal Nukak suicides, and dolphins gasping for air in the wake of the Gulf oil spill remind us that extinction involves more than physical loss. A soul does not die alone (Survival International, 2010; Bradshaw, 2012b).

The cure for society's malaise, wildlife extinctions, and environmental collapse will not come, as many argue, by *reconnecting*. One of ecopsychology's pioneers, Robert Greenway (2009), points out that this goal is based on an implicit misconception because "the existence of a relationship implies separate entities or processes." Congruent with a number of other cultures, this premise maintains that humans and their minds have never been disconnected from nature, and thus to imagine so is artificial. Greenway continues,

Reconnections between mind and nature . . . such as gardening, diet, natural dwellings, nature study, . . . transformation via vision quests and long-term wilderness immersion, though often beneficial (usually pleasurable as well) are still based on the illusion—the initial distortion—that *minds* can be separate from *nature*.

Greenway's observation underscores something key—namely, that the present planetary state that we seek to heal has been achieved by *denying connection*. Western society's false sense of separation has led to ethical distancing and legitimized an existence of psychological dissociation, "normal."

Now when learning of Harp seal mothers who have no ice on which to give birth and garbage islands the size of states floating in the Pacific Ocean, we understand that their plight is intimately linked with the hand that tosses away the plastic bag. When we bulldoze the earth to build a house, bears, deer, and squirrels watch *their* homes disappear under concrete. Their trees, berries, and bushes are appropriated with fence construction. We see wildlife through windows, thrilled to be so close, to feel *connected to nature*. Yet this yearning has not brought us closer after all. Proximity has been bought through violent assault. Suddenly, action made possible through psychological dissociation is no longer anonymous or blameless. Humanity is not afforded its past mental and ethical refuge from the consequences of its actions. Global collapse is personal.

Conclusion

A Yup'ik Eskimo handed me a scrap of paper whereon was penciled, "I am a Puffin!" . . . Here was a man who effortlessly negotiated the porous, wafer-thin membrane separating Homo from the Other. . . . Still alive. . . . Standing before me. . . . Symmetrical, convergent consciousness: the world before.
—Calvin Luther Martin, foreword to *Elephants on the Edge*

The lesson from the wild: life is only possible when lived in relational health. Saving wilderness comprises a radical conceptual and ethical transformation to relational living with nature, a trans-species paradigm that

> requires us to not only divest ourselves of the ethical double standard of the animal model in science and education, but transform how we deal with other animals in all aspects of living from a model of exclusivity to one of inclusion. This means finding ways to live that allow, to the best of our ability, all to thrive. (Marino, 2011)

Only then will our daemons revive or have the possibility of revival, and will we, as a species, *live*. Only then will we hear the mountains speak and thunder once again become an angry god.

This is not an easy process. Humanity is compelled to crack open the shell in which we have hidden:

> The shell is our fixed and sometimes unacknowledged assumptions and values. The shell has hardened from the influences we absorb consciously and unconsciously, from our language, our culture, and our personal needs and ambitions. These influences solidify and become our reality. To shed the shell, to view the world from a fresh perspective, we must risk turning upside down the most fundamental constructs of the past 10,000 years. To dream anew, we must see the shadows in all that we assume to be light. (Lawlor, 1991, p. 111)

There is a reason that we continue to exist in the way that we do—with more wars, more weapons, more killing of wildlife, and more denial that the condition of our society is as violent and angst ridden as it is—and the reason relates to Ernst Becker's proposition. The psychological mechanisms that permit a strident claim of love for nature to exist side by side with violence against wild flora and fauna express deep archetypal conflicts precipitated by a crisis in human identity—an identity, Becker (1997) maintains, that is constructed by the denial of death. The crisis in ego-human identity manifests in violent eruptions such as those that have marked genocides of indigenous people and animals. "The uprush of dark forces deployed itself in the individual . . . eventually breaking through the individual's moral and intellectual self control . . . [and] there was often terrific suffering and destruction" (Jung, 1946/1970, p. 242). This psychological crisis translates to the righteous battle for land and space, leaving elephants, tigers, and other wildlife as refugees in their own homes.

Nonetheless, if wilderness is to live inside and out, if we are to move from the isolation of survival to the inclusivity of life, then we must be willing to relinquish the protective shield of Cartesian privilege. The question is, Will we let down our accidental fences, make ourselves vulnerable in the nakedness of unity, and join life?

Author Note

This work is dedicated to O. Mein Gans for life. Parts originally published in *Vital Signs: Psychological Responses to Ecological Crisis* (Karnac Books, 2011; ISBN 9781780490489). Republished with the permission of Karnac Books Ltd. Finally, a warm thank-you to the editors for their patience and detailed attention to the mots justes.

References

Animals: Black Diamond. (1929, October 28). *Time Magazine*. Retrieved from http://www.time.com/time/magazine/article/0,9171,752326,00.html.

Becker, E. (1997). *The denial of death*. Berkeley, CA: Free Press.

Bogus, S. (1991). Mom de plume. In P. Bell-Scott (Ed.), *Double stitch: Black women write about mothers and daughters*. Boston, MA: Beacon Press.

Black Diamond at AI G. Barnes circus. (2011). *Elephant encyclopedia*. Retrieved from http://www.elephant.se/database2.php?elephant_id=2302.

Bradshaw, G. A. (2005). *Elephant trauma and recovery: From human violence to trans-species psychology*. (Unpublished doctoral dissertation). Pacifica Graduate Institute, Carpinteria, CA.

Bradshaw, G. A. (2009). *Elephants on the edge: What animals teach us about humanity*. New Haven, CT: Yale University Press.

Bradshaw, G. A. (2010). We, matata: Bicultural living amongst apes. *Spring: A Journal of Archetype and Culture, 83*, 161–183.

Bradshaw, G. A. (2012a). Can science progress to a revitalized past? In D. Narvaez, J. Panksepp, A. Schore, & T. Gleason (Eds.), *Human nature, early experience, and the environment of evolutionary adaptedness*. New York, NY: Oxford University Press.

Bradshaw, G. A. (2012b). Restoring our daemons. In M.-J. Rust & N. Totton (Ed.), *Vital signs: Psychological responses to ecological crisis* (pp. 89–103). London: Karnac Publications.

Bradshaw, G. A., Capaldo, T., Lindner, L., & Grow, G. (2008). Building an inner sanctuary: Trauma-induced symptoms in non-human great apes. *Journal of Trauma & Dissociation, 9*(1), 9–34.

Bradshaw, G. A., Capaldo, T., Lindner, L., & Grow, G. (2009). Developmental context effects on bicultural post-trauma self repair in Chimpanzees. *Developmental Psychology, 45*, 1376–1388.

Bradshaw, G. A., & Sapolsky, R. M. (2006). Mirror, mirror. *American Scientist, 94*(6), 487–489.

Buckley, C. & Bradshaw, G. A. (2010). The art of cultural brokerage: Re-creating the elephant–human relationship and community. *Spring: A Journal of Archetype and Culture*, *83*, 35–59.

Candland, D. (1995). *Feral children and clever animals*. Oxford, UK: Oxford University Press.

Capaldo, T., & Bradshaw, G. A. (2011). *The bioethics of great ape well-being: Psychiatric injury and duty of care*. Ann Arbor, MI: Animals and Society Institute.

Chomsky, N. (2003). *Hegemony or survival: America's quest for global dominance*. New York, NY: Metropolitan Books.

Clubb, R., Rowcliffe, M., Lee, P., Mar, K. U., Moss, C., & Mason, G. (2008). Compromised survivorship in zoo elephants. *Science*, *322*(5908), 1649.

Cohen, L. (2001). *God is alive: Magic is afoot*. Toronto, Canada: Stoddart.

dopetype. (2008). Golden compass "meet your own daemon" test. Retrieved from http://dopetype.wordpress.com/2008/11/22/the-golden-compass-deamon-test.

Dreier, T. (1913). *The vagabond trail*. Boston, MA: Griffith-Stillings Press.

Frankl, V. E. (1984). *Man's search for meaning: An introduction to logotherapy*. New York NY: Simon & Shuster. (Original work published 1946).

Freilander, A., & Landau, J. (1999). *Out of the whirlwind: A reader of holocaust*. New York, NY: Urj Press.

Greenfield, P., Keller, H., Fugligni, A., & Maynard, A. (2003). Cultural pathways through universal development. *Annual Review of Psychology*, *54*, 461–490.

Greenway, R. (2009). Ecotherapy news: Spring 2009. Retrieved from http://thoughtoffering.blogs.com/ecotherapy/2009/06/the-ecotherapy-newsletter-spring-2009healing-our-relationship-with-nature-ecopsychology-in-action-psychotherapy-as.html.

Herman, J. (1997). *Trauma and recovery*. New York, NY: Basic Books.

James, W. (1986). Confidences of a psychical researcher. In F. Y. Burkhardt (Ed.), *Essays in Psychical Research*. Cambridge, MA: Harvard University Press.

Judt, T. (1999, May 29). The courage of the elementary. *New York Review of Books*.

Jung, C. G. (1964). *Man and his symbols*. London, UK: Aldus Books.

Jung, C. G. (1970). The fight with the shadow. In R.F.C. Hull (Trans.), *The collected works of C. G. Jung* (Vol. 10). Princeton, NJ: Princeton University Press. (Original work published 1946).

King, B. (2009). Questioning elephants on the edge. Retrieved from http://www.bookslut.com/features/2009_11_015336.php.

Kirmayer, L. (2007). Psychotherapy and the cultural concept of the person. *Transcultural Psychiatry*, *44*, 232–257.

Klein, S. (2011, May 20). Seattle's Woodland Park Zoo defends treatment of elephants. Retrieved from http://mynorthwest.com/?nid=11&sid=484754.

Lawlor, R. (1991). *Voices of the first day: Awakening in the aboriginal dreamtime*. Rochester, VT: Inner Traditions.

Levi, P. (1995). *Moments of reprieve: A memoir of Auschwitz.* London, UK: Penguin.

Marino, L. (2011). A trans-species perspective on nature. Retrieved from http://onthehuman.org/2010/11/trans-species-perspective/.

Martin, C. L. (2009). Foreword. In G. A. Bradshaw (Ed.), *Elephants on the edge: What animals teach us about humanity* (p. ix). New Haven, CT: Yale University Press.

Narvaez, D., & Gleason, T. (2012). Developmental optimization. In D. Narvaez, J. Panksepp, A. Schore, & T. Gleason (Eds.), *Human nature, early experience, and the environment of evolutionary adaptedness.* New York, NY: Oxford University Press.

Noë, A. (2008). The life is the way the animal is in the world: A talk with Alva Noë. *Edge: The third culture.* Retrieved from http://www.edge.org/3rd_culture/noe08/noe08_index.html.

Noë, A. (2009). *Out of our heads: Why you are not your brain, and other lessons from the biology of consciousness.* New York, NY: Hill and Wang.

Paulson, T. (2005, March 1). Inseminating elephant takes 2 Germans, an ultrasound, and a very long wait. *Seattle Post-Intelligencer.* Retrieved from http://www.seattlepi.com/default/article/Inseminating-elephant-takes-2-Germans-an-1167610.php.

PhinneyWood.com. (2011). Zoo tries again to artificially inseminate elephant Chai. Retrieved from http://www.phinneywood.com/2011/06/09/zoo-tries-again-to-artificially-inseminate-elephant-chai.

Pullman, P. (2003). *The golden compass (his dark materials, book 1).* New York, NY: Laurel Leaf.

RoadsideAmerica.com. (n.d.). Two heads of Black Diamond. Retrieved from http://www.roadsideamerica.com/story/7832.

Savage-Rumbaugh, S., Wamba, K., Wamba, P., & Wamba, N. (2007). Welfare of apes in captive environments: Comments on, and by, a specific group of apes. *Journal of Applied Animal Welfare, 10*(1), 7–19.

Schore, A. N. (2005). Attachment, affect regulation, and the developing right brain: Linking developmental neuroscience to pediatrics. *Pediatrics in Review, 26,* 204–211.

Survival International. (2010). *Progress can kill.* Retrieved from http://www.survivalinternational.org.

YouTube. (2011a). *How to masturbate an elephant* [Video file]. Retrieved from http://www.youtube.com/watch?v=FX9Fc2aZSkc&feature=related.

YouTube. (2011b). *Orca semen collection* [Video file]. Retrieved from http://www.youtube.com/watch?v=TIU2-m_Vc7U.

7

A Wild Psychology

Ian McCallum

I know what I want to do for the rest of my life. I would like to be a voice, however insignificant, for the wild animals of the world, the wild places, and most important, the wild part of the human psyche. This decision is born out of my concern for the environmental issues of our time, especially the human contribution to our present state of affairs. We are in trouble. Our home is a mess. In our search for reasons, it would appear that we have failed to grasp the reality of our timing, place, and balance in the fragile web of life.

As human population growth continues along its accelerated path, there is no escaping the fact that our present lifestyles and material aspirations are undermining the conditions for human life itself, not only the biological conditions—clean water, clean air, and sustainable soils—but also the conditions for our psychological well-being. With the world's forests, wetlands, and peat beds under increasing threat—50 percent of these areas have been destroyed by human beings over the past hundred years, but they have not vanished completely—we are in danger of losing what I believe to be the wild landscapes of the soul. To me, these landscapes are a critical part of human identity. They nourish, often unconsciously, one of our greatest needs: the need to belong. If this is so, then surely it is simply not enough that we champion the protection of these endangered regions, for which, thankfully, there are many committed individuals worldwide doing just that. Our task, I believe is to carry this kind of commitment to another level. We need to be protecting the conditions for human sanity and, dare I say it, human life itself.

To do whatever I can as an individual to address and perhaps make a small contribution toward the redress of the disconnection between humankind and nature, I have become increasingly involved in environmental education, but with a fairly specific focus—conservation and leadership issues. My work as an analyst is about leadership, although not in the usual understanding of the word. I am not particularly interested in helping people to become successful executives. Instead, I try to help my patients become "captains of their own ships," be comfortable in their own skins, own up to their own contributions toward their personal problems, and learn that individuality is impossible outside of relationships, and that these relationships are not confined to human beings.

Extending my work beyond the consulting room to the outdoors and my education portfolio with the Wilderness Foundation, I have been deeply impressed by the passion, intellect, and eloquence of individuals who are committed to doing something positive about the various environmental issues of our time. I am amazed, but not surprised, that one burning question regarding the human contribution to our predicament is seldom, if ever, raised: Why as a species do we behave in such a destructive way in the first place? In keeping with my analytic work, this question is closely related to owning up to our personal contributions to our suffering. Beyond the consulting room, the answer to this query is slow in coming, and the reason for this, as I see it, is that we don't really want to know. The answer, it would appear, is too painful, too overwhelming, and possibly too debilitating to confront. We don't want to face up to the wild, animal, and evolutionary origins of human behavior. We tend not only to turn a blind eye to what is staring us in the face but attack it as well. We have confused wildness with savagery. Anything wild, spontaneous, or indifferent to human existence tends to be interpreted as either threatening or alien. The result of this distancing is best summed up by the ancient observation, "What we don't understand, we fear. What we fear, we ultimately destroy."

As a psychiatrist, I want to know why we behave the way we do. In search of answers, I have turned to nature and its survival-oriented lessons, and inevitably, to evolutionary thinking and Charles Darwin's principles of natural selection—variety, heredity, and common ancestry. It makes sense to me that the thread that binds us to all living things is a wild one. It is what this chapter is about.

I think we all know, deep down, that we have to live differently, but to do this, we first have to think differently. Clearly, we are looking at the need for a radical change in our attitude toward ourselves as well as our environment. The word radical comes from the Latin word *radicalis*, pertaining to the root . . . going to the root . . . fundamental. Applying this definition to the task of thinking differently, it is crucial that we investigate and understand how and why we lost our balance in the first place. We have to get to the root of our problem. We need to rediscover the essentials of a wild but human psychology.

Strictly speaking, there is no such thing as human nature. There is only nature and the very human expression of it. To understand this view of nature is to appreciate the significance of what we need to do if we are to restore not only the lost balance but also our position as a privileged human animal nature. It is to understand these words from the US poet Walt Whitman in his poem, *Song of Myself*:

I think I could turn and live with the animals . . .
they show their relations to me and I accept them,
they bring me tokens of myself, they evince them plainly in their possession.
I wonder where they get those tokens,
Did I pass that way huge times ago and negligently drop them?
(cited in Bloom, 2004, p. 542)

To me, the human psyche is alive with the tokens of the wild. And yes, I think we have been negligent. We dropped these tokens. Fueled by a lopsided sense of position, entitlement, and dominance over everything else on our planet, we carelessly dropped these tokens about ourselves from the list of the conditions for human survival. We have gradually but progressively lost track of our animal nature and what wildness really means. Simply because of its evolutionary and therefore survival significance, to my mind it is crucial that we rediscover these wild tokens in ourselves; that we be willing to understand how they can be tracked and appropriately translated at the biological, social, and psychological levels of human existence, and that we protect them . . . fiercely.

A Wild Psychology

The first step in the cultivation of a wild psychology, and with it, a practical redress of restoring our lost balance with nature, is to own up to the origins and possible direction of our evolutionary spoor. In short, we have to take charge of our evolution—not our biological evolution, for this is not our primary concern. And even if we could, what precisely would we want to change, and in what order? Rather, we can begin to take charge of our psychological evolution. We can, each one of us, learn to take responsibility for, accept, and harness the discarded animal tokens of our individuality—and more, our humanity. It is crucial that we learn to shift our focus from self-centered human interests to that of the web of life along with the *relationships* (human and nonhuman) on which our existence as individuals and societies ultimately depends. It is important that we therefore understand, accept, and acknowledge the hardwired, emotion-charged instincts and motivational drives of human behavior. Without them, we would not survive.

Remember, to control human behavior is one thing. This is the easy part for it does not require any depth of understanding. Legislation, social pressures, force, and the promise of punishment are often all that is needed. Clearly, these strategies have not been that effective. To understand human behavior, on the other hand, is a different demand altogether. Like the transformation of fire, the original raw flames of warmth, into heat and electric energy, you first have to understand human behavior along with its origins, motives, potential, limitations, and power. How else, without an understanding of where we have come from, are we to appreciate the nature of our biological drives, and what they mean as well as the different ways they are hidden, disguised, and expressed in every human being? How else, without understanding, are we to effectively say yes and no to what is deepest and most ancient in us—the sociobiological forces of territory, ownership, belonging, kinship recognition, and their dark sides: xenophobia, racism, fundamentalism, and war? We preach democracy and equality of all human beings, but why do discrimination, authoritarianism, tyranny, fundamentalism, and power struggles persist?

In spite of the growing interest in evolutionary biology, it would appear that on a psychological or, if you prefer, personal level, we struggle to own up to our animal ancestry. Somehow we are above all that. We just don't want to know. For fear of what we might find out about ourselves as well as about our true status and that of all the living things we have historically regarded as subservient to our existence, we refuse to take seriously the evidence of our less sophisticated animal origins. Darwin thought otherwise.

The Wild Psychologist

A brave and intellectually honest person, Darwin was a psychologist in his own right. I believe we are indebted to him for his insights. It is worth remembering that in 1836, when Darwin set sail on the *Beagle* as the onboard naturalist, the prevailing theological interpretation of the origin of Earth was that it was formed intact, on October 23, 4004 BC. Creationism was the order of the day. As we shall see, the evolution of human beings, mind, and human behavior was fundamental to Darwin's system.

Today what we regard as psychology was in his time the subject of meta-physics—a branch of philosophical inquiry into the nature of how things work, or as the name suggests, what lies "behind" the laws of physics. "Puzzling about metaphysics," Darwin once said, "is like puzzling about astronomy without an understanding of mechanics" (quoted in Sulloway, 1980, p. 241). The same could be said for puzzling about psychology.

It is clear from his research that Darwin was intent on establishing a "stabil [sic] foundation" between evolution and psychology. His grasp of a continuity of humans and animals, for example, led him to investigate as well as correlate the evolution of intelligence, the nature and origin of instincts, and the expression of emotions in humans and animals. The latter was to become the title of one of his books. For Darwin, psychology was a mind–body phenomenon. He studied gestures, facial expressions, and nonverbal communication, noting, as any parent and pet owner knows, that "preverbal infants, like many animals, seem to understand spoken words even though they can't articulate them" (quoted in ibid., p. 245).

When his first child was born (he had ten, of whom eight survived), Darwin kept a diary of the child's first three years of mental and behavioral development, later comparing these with observations of his other children. He took his children to the zoo, and observed with surprise their fear of the larger and more exotic animals without any prior exposure or experience of these creatures. Darwin wrote, "May we not suspect that the vague but very real fears of childhood which are quite independent of prior experience are the inherited affects of real dangers and abject superstitions during ancient savage times?" (quoted in ibid., p. 244). In 1876, in what could be regarded as the first formal journal of psychology, *MIND*, he published his "A Biographical Sketch of an Infant."

Darwin was indeed a psychologist, and not only a child psychologist, for there were other essential psychological themes and dynamics that caught his attention. If you wanted to understand the present, for instance, it was important to understand the past. He was also interested in the development of sexuality—for example, the sexually exploratory behavior of children and animals. Darwin saw a link between the act of a dog smelling the vagina of a potential mate and that of the human acceptance of the odor of our own pudenda. As he put it, "The smell of our own pudenda is not entirely disagreeable" (quoted in ibid., p. 242). Another theme was that of adaptability and the survival significance of instincts—something that the Swiss psychiatrist Carl Gustav Jung would advance in his theory of archetypes, describing them as the psychological equivalents of our biological instincts. Adaptability, arguably the central

theme of any form of modern psychotherapy, was also central to the work
of the Swiss cognitive psychologist Jean Piaget. Yet it was Darwin's observations and interest in the psychological, sexual, and physical developmental stages of the child that should alert us to his influence on one,
particularly significant psychologist: Sigmund Freud.

Freud was initially an enthusiastic disciple of Lamarckian evolution
and the idea that adaptation was internally driven rather than subject to
natural selection, but he later swung toward Darwinian observations. It
comes through in his writings about the archaic nature of the unconscious
and infantile erotogenic zones along with his insights into conscious and
unconscious psychic adaptability. It was from these insights that he developed his theory of ego defense mechanisms, and lest we forget, it was
Freud who proclaimed that "the ego is not the master in its own house"
(quoted in ibid., p. 276)—an acknowledgment of the power of the unconscious or, if you prefer, our evolutionary mind. Toward the end of his life,
Freud recommended that the study of evolution be included in the training of every prospective psychoanalyst.

I would like to share with you the relevance of evolutionary thinking
to my work and what has followed—my commitment to being a voice
for the wild part of the human psyche. For a start, evolutionary thinking, because of its strong emphasis on biological connectivity, has left me
with no choice but to show a greater compassion for all living things. It
has forced on me the need to redefine who "I" am and who "we" are as a
species. In other words, where does the I and we of our identity as human
beings begin and end? Thanks to the unraveling of the human genome
and realization that we share a common language with every biological
species—the language of DNA—the poets have been proven right. We
are brothers and sisters. The web of life is a reality. That we are stardust,
as William Blake wrote in the nineteenth century, is a fact. The hydrogen
atoms in your and my bodies, in trees, rivers, and rain, are 13.7 billion
years old. Iron, the central ingredient of hemoglobin, comes from exploding stars, as does magnesium, potassium, and calcium. Without these elements, we would not survive. Carbon comes from the sun. Every mammal, from mice to honey badgers and elephants, has more than 90 percent
of our genome. You and I can take a blood transfusion from a chimpanzee with impunity. We are connected. We share a common ancestry. Be
kind therefore; we are all fighting a fierce battle . . . and the name of that
battle is "survival." If this is so, and I cannot counter this argument, then
evolutionary thinking invites us to not only acknowledge a common law

of survival but examine the way it translates at the biological, social, and psychological levels of human existence as well.

The Law of Survival

The law of biological survival is simple: *for an organism to survive, it has to exist in an element in which it can find or produce enough food to give it enough energy to live long enough to reproduce.* This law can be pictured as a vertical axis on which the basic requirements for biological survival are spelled out—territory (element), food (resources), energy, adaptability, time, and sex. There is also a horizontal axis, however, to what I now regard as a survival compass. This axis spells out the required (behavioral) skills—or if you prefer, the varied but necessary adaptive strategies—to establish territory, acquire food, sustain energy, stay alive, and successfully reproduce. It is balanced by two basic behavioral strategies: fight and flight. Moving along the spectrum from fight, on the one end, to flight, on the other, these strategies include challenge, competition, opportunism, risk taking, deception, cooperation, compromise, sacrifice, withdrawal, and escape.

Euphemistically referred to as the "Four Fs," the law is a function of the interactions between food, sex, flight, and fight.

If you wish, you can change the wording. You can replace the brutish-sounding fight and flight with approach versus avoidant behavior, or if you prefer, hide-and-seek. Food can read as resources and sex as reproduction. It doesn't matter. The law, as we will see, remains largely the same.

Applying the law to a human culture or any other human organization, the law will read something like this: *for any organization to survive, it has to have an established brand/identity/niche* (territory) *in which it can draw on or produce enough resources—such as capital, human resources, or food—to give it sufficient resilience* (energy) *to be sustainable* (time), *productive, and influential* (reproduction). The horizontal axis of the organization's survival compass includes the full spectrum of fight-and-flight (adaptive) strategies applicable to biological survival.

And the same holds true for psychological survival. To survive psychologically, the law would read similarly: *to survive psychologically, is to have a sense of self* (territory) *stable enough to effectively draw on, and utilize inner and outer resources—such as knowledge, information, meaning, or food—to keep sufficiently motivated* (energy), *effectively deal with*

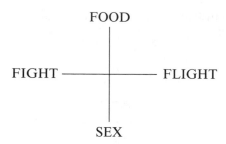

Figure 7.1

The law of biological survival

life's challenges (time), *and be creative* (reproduction). The horizontal axis is in principle no different.

I can see the raised, critical eyebrows. There's something missing. It can't be as simple as that, and you are right. Where do spirit, morality, and freewill feature in this formula? And what about play? I will deal, first, with play. Play is the missing "F" in this survival formula. It stands for fun, frolic, and fantasy. Without play, the lives of certain bird species and that of every mammal would be the poorer, if not impossible.

To play is central to the survival compass of every mammal—the essential, skill-refining activity for the establishment of turf, acquisition of food, and foreplay for sexuality and mate selection as well as the testing ground for the varied, adaptive strategies on the fight-and-flight axis. In social species, it is a prerequisite for the building, reinforcing, and restructuring of relationships, hierarchies, and alliances. In humans, play, problem solving, and imagination are inseparable. Play stretches our boundaries. It is the home and testing ground for the absurd and outrageous. From puzzles and mind games to boardroom brainstorming and strategy planning—what is sometimes referred to as *serious* play—the more dependent a species is on trial-and-error learning, apprenticeship, and mentorship in order to thrive, the greater the need for play.

Playgrounds are movable. They are in the mind as much as they are measured. Our physical playgrounds may be concrete or clay, but they remain our fields of dreams, our theaters for the shared sweetness of victory, and the learning curve that comes with the licking of wounds. Games, whether we win or lose them, prepare us for life, for as we all know, we can't and dare not win them all. Nobody would want to play with us. With obvious exceptions, play is nature's invitation to test one's physical

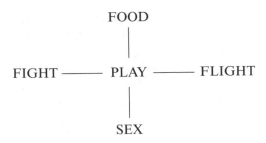

Figure 7.2
Play is the missing "F" in this survival formula, and it is where it is most needed in our survival compass: in the middle

and intellectual skills without malice, and I know it may come across as cynical, but think about it, play is never that far from being a dress rehearsal for the "kill"—for turf, status, food, the trophy, or the mate. Play is where it is most needed in our survival compass: in the middle.

Restoring the Balance

As we begin to address the various ways and means of restoring the lost balance between humankind and nature, I would like you to be mindful of the following three images. To me, they are relevant not only to the task but also to a personal sense of direction and hope. I want you to imagine a picture of Earth taken from deep space. The round, blue planet in and of your imagination is an image of home—the only home we have. There may well be other Earthlike planets somewhere in the cosmos, but as far as we are presently aware, Earth is the only place in the entire universe that the conditions for biological life as we know it prevail. Welcome home.

I now want you to imagine a diagram of the human brain divided into three distinct evolutionary components: a brain stem or reptilian component; a middle, limbic, or ancient mammalian component; and a cortical or modern mammalian component. Introduced by the neurobiologist Paul MacLean (1987) in the 1980s, this so-called triune brain is a powerful reminder of the likely evolution of the modern human brain. While there is no obvious clear delineation of our brains as described above, the concept of an interdependent yet emerging evolutionary brain is a powerful one. Today, with the unraveling of the human genome, and with it the evidence of our DNA relationship to all living things, I cannot think of a more plausible explanation of the origins and adaptations of what most defines us as a species: our ten-trillion-celled nervous system. Without it there would be no human intellect, psychology, or spirituality. Ultimately, this model draws attention to the likelihood that every nerve cell in our bodies originated at some stage in its evolutionary history from one, individual cell. It is a sobering thought. Is it too much to ask that we pay tribute to that ancestral cell, its early interaction with the environment, and subsequently, the evolved neurobiological underpinnings of human behavior?

The third image is that of a press photograph taken not far from Beijing's Tiananmen Square. The setting was, of all places, the Avenue of Peace. The year was 1989, and the event was the student revolt against an oppressively conservative Chinese government. In the picture, a student, sometimes referred to as "tank-man," dressed in a white shirt and dark long pants, his back to the camera, is standing in front of a column of four military tanks. The message is clear. In a moment of courage, madness, or as I read it, an ethical imperative, a student says "NO!" Standing his ground, a single human being brings the rolling tanks to a halt. As you can imagine, his action contributed to a massive and brutal clampdown by the Chinese officials on any form of antigovernment protest. I have no idea what happened to this brave dissident, but there are no other images I know of that quite captures the definition and loneliness of the individual hero as this one. Yes, it speaks of the power of the individual, of never underestimating the influence of individual life, yet to me it speaks of something else. I see it as a powerful challenge for each one of us. In this image, I see the metaphor of a thin and fragile neocortex standing up to the emotion-charged drives of the human brain stem and limbic systems.

Tied to these images are the names of individuals who have contributed to my present appreciation of the ancient, biological underpinnings of human behavior. I will focus on two names: MacLean, who I have already mentioned, and the US neuropsychologist Jaak Panksepp. There have indeed been others—for example, Richard Dawkins, who in spite of his quixotic tilts at both gods and religions, has been an inspiring voice for evolutionary thinking. Another influential voice is that of Harvard entomologist and sociobiologist E. O. Wilson. Wilson (1984, p. 134) coined the term for a phenomenon we all intrinsically know: *biophilia*, "the inevitable and innate human tendency to be drawn to life and lifelike forms, and at times to affiliate emotionally with these forms." And then there is the Russian biologist Theodosius Dobzhansky, who announced, "Nothing in biology makes sense except in the light of evolution" (quoted in

Pinker, 1999, p. 210). I will venture that nothing in psychology makes sense either . . . outside of that same light.

MacLean, as noted above, is responsible for the image of the triune brain. He may not have been the first person to come up with the idea, but as far as I am aware he was the first to portray it in such a profound though simple diagrammatic way. Clearly there is no absolute or measurable demarcation of the reptilian, paleomammalian, and neomammalian brains, but that is not the point. It is the concept that is important. This diagrammatic notion of emerging neurological substrates provided the scaffolding I needed for the thoughts and ideas I had been wrestling with for some time. I saw in the divisions a conceptual "home" in humans for the evolution of three survival-oriented intelligences: a *reactive* intelligence (reptilian), *responsive* intelligence (ancient mammalian), and *reflective* intelligence (neomammalian). Humans are examples of an interesting mixture of all three. I wish to stress that because the reflecting intelligence is the most recently evolved, it should not be read as the most significant or effective. This is not a hierarchy of intelligences. We need all three. Nevertheless, they each have their place, with each of them primed for specific situations.

Reactive Intelligence

As I see it, our reactive intelligence is mostly a brain stem or reptilian attribute. Characterized by attributes generated in the brain stem—being awake, alert, and aware—our reptilian intelligence is reactive, defensive, and explosive. Linked to the hypothalamus—the central region for the regulation of hormones and vegetative functions (such as thermoregulation, hunger, or thirst) in our brains—our reptilian brain stem is home to the hardwired survival- and emotion-charged circuits of lust, rage, fear, seeking, and panic. Panksepp (1998) elegantly defined and described these circuits as well as two others necessary for mammalian survival: nurturance and play.

Our reactive intelligence, then, is primal. It flies its flag in issues involving territoriality (establishing, protecting, and expanding territory), exploration (finding resources), camouflage (deception), ritual (the marking and patrolling of territory), repetition (game paths or hunting methods), reproduction (such as mate selection, nest building, or routine), and unconditional "self-survival." It is conceptually, the "me" part of the human brain. Sharing and altruism is not a reptilian trait. The brain stem wants it all, wants it now, and wants as much as possible. Hoarding, consumerism, and what is easy to describe, through the triune eye, in others as insatiable greed is driven by the brain stem. Impulsivity, then, is a core feature of a reactive intelligence—something we have all had to deal with in our own makeup.

As we know, it is easier to recognize reptilian behavior in others than it is to do so in ourselves. Our reptilian brain does not know the difference between constructive criticism and threat. It reacts to the immediate—flight or fight. It is not interested in open societies or democracy. In Freudian terminology, our reptilian brain is the id, the home and source of that which is most primal in us. What is more, it is alive and well. The crocodile is in and with us.

Responsive Intelligence

The responsive intelligence of the paleomammalian brain emerges from its reactive, reptilian roots. Home to the hardwired circuits of nurturance and play as defined by Panksepp, this evolved level of intelligence is necessary for social cohesion. Compared with reactivity, responsiveness implies a little more give-and-take. Characterized by a capacity to be aware of the awareness of the other, and in humans, to be aware of the needs of the other, it is a less explosive intelligence than its reactionary predecessor and a little less defensive. The relevance of a responsive intelligence is most obvious when considering the underlying conditions for social survival: nurturance, play, kinship recognition, hierarchies, sharing, prolonged care for the young, and in humans, codes of conduct and care for the elderly.

Our paleomammalian or social brain is the stomping grounds for what Freud called the superego, the "thou shalt and thou shalt not" region of human social functioning. Compared with the "me" orientation of our reptilian brain, our limbic/paleomammalian brain is a little more "you–me" oriented.

One of the most potent lessons I have learned from my understanding of the paleomammalian aspect of the evolutionary brain concerns the nature of kinship recognition. Kinship recognition is critical for social and cultural survival. It has a dark side, however. Supported by common languages, physical peculiarities, dialects, beliefs, ceremonies, membership, and purpose, kinship recognition is at the core of so-called in-groups. It therefore follows that nonmembers or nonkin are out-groups. There are, of course, many examples of harmless in-group/out-group rivalries, but the really dark side of this otherwise-natural bias looms large when the out-group, usually the minority, is perceived as the enemy. This happens when our reptilian brain overrules the responsive and reflective intelligence of the cortex. Sometimes described as the destructive projections of what in-groups fail to see in themselves, to me our territorial, reptilian shadow stands as the most likely explanation of the dynamics of racism and xenophobia.

Reflective Intelligence

The evolved human cortex otherwise described as the neomammalian brain can be seen to be the substrate for a reflective intelligence. Without it, sequencing, symbol formation, and effective situation analysis would be impossible. It is the favorite territory for the ego, the orchestrating principle of our waking reality. It is potentially the "I–thou" faculty of human behavior, the substrate for the integration not only of our different intelligences but the different levels of awareness generated by the "older" parts of our brains as well—being awake, alert, and aware.

In addition, though, it brings another two levels of awareness to the human equation. Both, I suggest, are critical for our survival. The first is the capacity for self-awareness—a *conscious* sense of "me" and "not me." This is a difficult one for it may not be restricted to human brains. For instance, if a leopard chooses to rest on a sturdy branch of a tree compared with a more flimsy one, could we say that the animal is aware of its own weight when picking a branch that can hold it or not? If it is a conscious decision, then we would have to conclude that the animal is self-aware. If not, then we have to agree that there is at least an intrinsic capacity in the animal to measure itself—such as its weight, size, and strength against something else in its world. Could this intrinsic ability be the necessary foundation for conscious self-awareness?

As far as I know, there are at least three nonhuman animals capable of conscious self-awareness: chimpanzees, dolphins, and elephants. How do we know? One way to measure self-awareness is to do the "mirror test." "That's me"—or for some of us, "that can't possibly be me"—is what most humans conclude when they see themselves in a photograph or mirror. Put a mirror in front of an animal, and if it recognizes itself, then we can say it is self-aware. In chimps and elephants, the mirror test involves placing a blob of paint on the animal's face. In most cases, but not all, it does not take long before the animal sees the blob of paint, recognizes it as artifact, and then begins completing the task of removing it. Dolphins, on the other hand, test a little differently. In one test case, after numerous encounters with its image in a mirror, a "performing" dolphin plucked a floating leaf from the surface of the pool in which it was being kept. The dolphin carried the leaf to the mirror, and on seeing the leaf in its mouth, dropped it and swam off. This action probably confirms that the dolphin understood the relevance of the leaf as separate from itself.

The second, additional level of awareness required of a reflective intelligence is the capacity *to be aware that you are aware that I am aware that you are aware.* This fourfold interchange of awareness is a critical aspect of the evolved human game of hide-and-seek. It keeps us on our toes. For instance, "I wonder if you are aware that I am aware that you are aware that I am aware that I should be bringing this chapter to a conclusion?" This internal dialogue or interchange appears to be uniquely human. It is how, as members of the human social species, we keep a check on each other, and especially how we keep a check on how others are keeping a check on us. We would not be able to do this without the neomammalian cortex.

It is important to remember that reptiles are not purely brain stem creatures. Like all mammals, they have limbic systems as well as cortical gray matter; it's just that these structures are less anatomically and physiologically evolved compared with the brains of birds and mammals. In other words, let's not forget that reptilian and other wild animal behavior is not entirely determined by brain stem instincts. Learned behavior, or what could be called "top-down" influence, becomes more significant the higher up we exist on the triune ladder of intelligence. As we know, a crocodile can also learn a few things. They are indeed intelligent, or as intelligent as they need to be in their particular niche on Earth.

Yet there is another aspect of the reflective neomammalian brain that needs to be highlighted, and that is relevant to this chapter, our survival, and the challenge to understand the wild forces of human behavior: without the neurological substrate for a reflecting intelligence, we would be unable to *inhibit,* modify, restrain, or transform the intensity of the underlying emotion-charged circuits that emanate from the older, primal parts of our brains. Inhibition—and I am not speaking about the unconscious ego defense mechanisms of denial (such as repression or sublimation)—is about the deliberate or conscious capacity to say no. Hence, the image of the student standing in front of the rolling tanks in Beijing in 1989. I see this student as a heroic representation of what our rather fragile and recently evolved neomammalian brain is capable of: saying yes and no to what is deepest in us. The tanks are a representation of the human limbic system and brain stem. The student is a representation of a fragile human cortex standing its ground, saying no not only to the tyranny of dictatorships and fundamentalism but also the ancient biological and social drives of our own human tendencies. The tanks are not going to go away; still, they have been brought to a halt, at least for a while. To me, the words yes and no are the most powerful ones in the vocabulary of a species that is capable of deciding what to do about its future. To say no to the tanks is to say yes to something else—ultimately, to a deep and *personal* ethic.

Amorality, Morality, and Ethics

Viewed through a triune eye, our reptilian brain is amoral. It has no consciousness of right and wrong, good and bad. Reptiles do what they do without conscience—with no remorse or regrets. They don't take anything personally.

Our social or paleomammalian brain is different. Care, fair play, and sharing have their roots in this neurological bracket. In human societies, as we know, behavior that is good for the group and that which is not are both closely monitored. The same goes for baboons and chimpanzees. The England-based analytic psychologist Julian David (2007) reminds us that morality is a collective concern. The word moral comes from the Latin word mores—the customs of society and rules by which society works. It is about obedience.

Ethics is something different. Rooted in the neomammalian cortex, it is a reflective phenomenon. To me, ethics is about the individual rather than the group, and because it is based on personal experience, it is therefore a personal as opposed to collective code of conduct.

The word *ethic* comes from the Greek *ethos*—the spirit by which things live, the source of which is not culture but instead nature. The difference between morals and ethics, says David, is that one is cast in stone, and the other is spontaneous. He writes, "Ethics comes to each situation fresh, as if it had never happened before. . . . [I]t operates in a doubt-filled twilight with only itself to distinguish between what is more valuable and what less is and must be sacrificed. . . . [S]ometimes we must be able to go against the *mores*" (ibid., p. 36). Morality, then, is about social control—obedience to rules and regulations that are deemed to be in the best interests of society. Morality, how we ought and ought not to behave, is the language of the superego. Ethics, on the other hand, because it is both individual and personal, is about knowing when to say yes and no to the collective, but also to the emotion-charged psychobiological "tanks" of your own being. If morality is about the control of human behavior, then ethics is about understanding it—on a deeply personal level. If the moral stance was to crush the student uprising in Beijing, then the ethical stance was to say no to the flawed morality behind the reason for the rolling tanks.

Let's be clear about this: morals and ethics are not necessarily mutually exclusive. Instead, there is a natural tension between the two. And so it should be, for it is in this tension that the subject of freewill comes alive. Freewill is frequently regarded as one of the defining qualifications of humanity. Usually lumped into the often-meaningless bracket of human rights along with the freedom to say and do whatever we please, it is anything but. Instead, and I want you to think about this carefully, it is the freedom to say no. In evolutionary terminology, the freedom to say no" is a description of the inhibitory function of the reflective neomammalian cortex. Without the cortex, there would be no brakes, filters, or defenses against our individualistic limbic and brain stem demands. Without a cortex, there would be no such thing as a personal ethic. There would no ability to grasp the significance of the paradox where saying no to something is at the same time saying yes to something else. Without a reflecting cortex, the understanding of human behavior would be impossible.

Finally, what is the lesson in this call for a wild psychology? To me, and most important, it is this: to attempt to explain the underlying evolutionary substrates of human behavior in an evolutionary light is not to condone human behavior. It is to *understand* it. It is to be mindful yet realistic about our aspirations for the human species. Remember, the triune concept and associated intelligences is an *idea*, and if it helps us rediscover our relationship with the wild and the underlying forces of human behavior, it could be the idea that we most need to cultivate and protect. Evolutionary thinking has helped me see biological, social, and individual life differently . . . to be a little less surprised about the guiles and gumption of human behavior. For example, power struggles, territorial claims, in-groups, out-groups, and their dark sides—tyranny racism, ethnic bias, and xenophobia—will never go away. They will not spontaneously vanish. They have been around for a long, long time, and our reptilian and paleomammalian brains are not going to let them go. Our task is to understand the dynamics so that we can say no to them. This cannot effectively happen until we are aware of their evolutionary significance and, in this instance, neurological substrates.

Piecing all of this together, there is good reason to ask the following questions: Is evolutionary thinking nothing more than an intriguing or novel idea, another discardable, deterministic hypothesis on human life, or is it a manner of thinking that we should not ignore? I choose the latter. Evolutionary thinking is now an important part of my work. To see psychopathologies such as depression, phobias, adjustment, and anxiety disorders in an evolutionary light is to see them differently. To see maladaptive traits and behaviors as an exaggeration of what was once adaptive, or as Darwin put it, "the inherited affects of real dangers and abject superstitions during ancient savage times" (quoted in Sulloway, 1980, p. 245), goes a long way toward removing the sting of shame and helplessness from a patient's suffering. Evolutionary thinking, because of its profound ecological message, has steered me toward attending not only to the wounded interiority of my patients but also with time, to the often-unacknowledged exteriority of their lives—the streams and trees, landscapes and animals, in their lives, and ultimately, that wild thread that binds all biological life. As crucial as the interiority of one's psychological wounds may be, in my view, psychological healing will never be complete as long as it is seen to be separate from the world and web of life (McCallum, 2005; Roszak, Gomes, & Kanner, 1995).

To me, the human psyche is a body-mind-Earth phenomenon, an incredible interplay of genes, imagination, and biosphere. I can no longer escape the awesome biological underpinnings of human nature, that we are part of the processes of Earth and indeed the universe. To become aware of this connection and then to begin to live it is to add depth to the meaning of the word belonging. It is to become aware of a greater sense of synchronicity in our relationships—human and nonhuman. It will certainly go a long way toward a more practical, genuine reappraisal of our place and position on Earth.

Conclusion

To understand the wild history of human behavior, to accept and then consciously harness it, is going to be a huge, if not impossible, task. To attempt it is to be mindful of Seamus Heaney's comment about poetry: "Poetry always has the right to reach for the stars. The question is whether the poet can rise to that challenge" (quoted in O'Driscoll, 2008, p. 196). I don't think we have any choice. We are, after all, the reflecting "human animal." What a privilege. What a responsibility. Be kind therefore; we are all fighting a fierce battle.

Author Note

The theme and contents of this contribution to *The Rediscovery of the Wild* is based on a public lecture delivered in Cape Town in February 2010.

References

Bloom, H. (2004). *The best poems of the English language*. New York, NY: HarperCollins.

David, J. (2007). Morality and ethic: The question of conscience. *Mantis, 19*(2), 36.

MacLean, P. (1987). The triune brain. In G. Alderman (Ed.), *Encyclopedia of neuroscience* (pp. 1235–1237). Cambridge, MA: Birkham.

McCallum, I. (2005). *Ecological intelligence: Rediscovering ourselves in Nature*. Cape Town, South Africa: Africa Geographic.

O'Driscoll, D. (2008). *Stepping stones: Interviews with Seamus Heaney*. London, UK: Faber and Faber.

Panksepp, J. (1998). *Affective neuroscience: The foundations of human and animal emotions*. Oxford, UK: Oxford University Press.

Pinker, S. (1999). *How the mind works*. New York: Penguin Books.

Roszak, T., Gomes, M. E., & Kanner, A. D. (1995). *Ecopsychology: Restoring the earth, healing the mind*. San Francisco, CA: Sierra Club Books.

Sulloway, F. J. (1980). *Freud, biologist of the mind: Beyond the psychoanalytic legend*. New York, NY: Fontana Books.

Wilson, E. O. (1984). *Biophilia: The human bond with other species*. Cambridge, MA: Harvard University Press.

8

Culture and the Wild

E. N. Anderson

Though you think sweet, yonder in your church, the gentle talk of your students, sweeter I think the splendid talking the wolves make in Glenn mBolcáin.

Though you like the fat and meat which are eaten in the drinking-halls, I like better to eat a head of clean water-cress in a place without sorrow.

—Mad Sweeney, quoted in Jackson, 1971

All cultures worldwide must necessarily confront the contrast between cultural management and representation of the land and the lack there-of—between more "cultured" and more "natural" environments (Lévi-Strauss, 1964). Some contrast the human realm with the wilderness; others have quite different ways of conceptualizing the contrast. Differences in cultural views are associated with differences in actual management. Some cultural groups hardly affect the landscape at all. Others work to manage the natural landscape for the sustainable use of resources. Still others completely transform large sections of the land, but leave other sectors less affected, either for resource use or recreation and spiritual renewal. These numerous and varied ways of constructing the wild, and contrasting it with the tamed, can inform our future use of the world. Studying them and teaching their generally applicable conclusions therefore might be a target for future efforts.

The first half of this chapter deals with different concepts related to the Western ideas of "wild" and "tame." Native Americans, medieval Irish, Chinese, and other East Asians all have quite different ideas about this opposition, as will become apparent below. The chapter continues with the ways these ideas relate to actual management, especially sustainable management. Often this is done through the religious construction of a transhuman society and religious rules of management, which leads to many cross-cultural equivalents of the "dark green religion" recently described by Bron Taylor (2010). Finally, I provide some applications, including alternatives to displacing indigenous peoples in order to create reserves, setting new social priorities, educating for sustainability, and seeing all lives as persons, or at least in some sense subjects deserving of consideration.

Cultural Ideas of the Wild

The idea of wild nature opposed to humanized space is an old one in the West, beginning long before the Romans gave us the word *natura*.[1] In the Western world, it has changed over time, as have the values of the wild.

The first stereotypical opposition of wild and tame in recorded history lies in the *Epic of Gilgamesh*. Here Gilgamesh, the model of the sophisticated, cultured urbanite, finds his opposite number and beloved partner in Enkidu, the wild man:

His whole body was shaggy with hair. . . .
He knew neither people nor settled living,
But wore a garment like Sumukan [the god of wild animals, clothed in skins].
He ate grasses with the gazelles,
And jostled at the watering hole with the animals

(cited in Kovacs, 1985, p. 6)

Like savages in fiction ever since (Bartra, 1994, 1997), Enkidu used a club, was powerful and lusty, and succumbed all too easily to alcohol. The pairing of clever hero and uncouth but strong foil is with us yet, especially in movies and comics (as, for instance, in the French comics by René Goscinny and Albert Uderzo featuring Asterix and Obelix).

In ancient Mesopotamia, cities and states began their career on Earth by diking, draining, and cultivating the entire landscape, and turning it into a realm of wheat, barley, sheep, goats, and little else. This started the Western regions on a career of extensive farming of a few commodities, and its career of regarding the wild—and "wild men" like Enkidu—as outside the pale. Destroying nature for monocrop agriculture thus is not a modern, capitalist, or "developed" condition but rather a basis of life in the Western world for five thousand years.

The Bible notoriously gives "man" full "dominion" over nature, but only in the first chapter of Genesis. The second chapter contains an entirely different creation story, which makes people the stewards of God's landscape, not its dominators. This split in attitudes runs through the entire Bible. Usually the stewardship view prevails (Hessel & Reuther, 2000; Tirosh-Samuelson, 2002). As such, we have the stunningly beautiful nature poetry of the Song of Solomon and the Psalms. On the other hand, the horrors of "wilderness" are graphically described in many places. The

Book of Isaiah includes both lyric descriptions of peaceful nature and horrific descriptions of wilderness. The latter, however, probably refer to the results of human mismanagement—they are punishments for sin, including the sin of taking poor care of one's irrigation works; the wilderness is a salt marsh, probably the result of irrigation mismanagement. The Bible mentions rather few plants and animals, and describes an agriculture based on only a few crops. The Chinese *Book of Songs* (compiled circa 500 BC) mentions many more plant species, yet is only the length of one of the longer books of the Bible.

The ancient Romans saw the forest as the ultimate wild, and gave us the word *silvaticus*, surviving, with increasingly negative connotations, in Spanish as *selvatico* and *salvaje*, in French as *sauvage*, and in English as *savage*. Latin also had the word *ferus*, "wild" in general; it may be distantly related to our Germanic word wild. Our wilderness is a place of wild animals—"wild-deer-ness" (or the word may be just a variant of wildness). The word wild has two meanings in English and related languages: "natural" and "violently crazy." The *Oxford English Dictionary* informs us that both meanings have been implied by the word as long as its use can be traced.

Other languages divide the world in quite different ways. The Yucatec Maya with whom I work in southern Mexico have no equivalent to the word natural, and none with the unpleasant connotations of our wild (Anderson, 2005). Their best approximation is strikingly close to the Latin *k'aax*, "forest" (or *k'aaxil*, "of the forest"). The difference is that the Maya manage their forests. They cut them on fairly regular cycles to make fields and then allow the forests to regrow, so that even an old forest is known to be former farmland. They manage their forests, continually pruning, trimming, harvesting wild fruits and herbs, hunting game, and taking logs and poles for construction. Many forests are selectively enriched with useful trees because these are saved when fields are cleared or even planted in the fields. The result is that all parts of the landscape are managed space, and no wild or wilderness land exists. Even when they speak Spanish, the Maya do not normally use words or images that would make nature wild. A wild animal is *bravo* or *arisco*—words that refer to uncontrolled and untamed behavior; these terms apply to a refractory horse or cow, but conversely, are not used for forest animals that are peaceable and tranquil. The forests that are least actively managed, and that we would call wild, are loved. "The forests enchant me, the trees enchant me," as my Maya friend and coworker Felix Medina Tzuc once said to me.

The traditional comprehensive and holistic land management system was maintained as well as sanctioned by indigenous Maya religion. The Maya have moved into the Christian orbit since the Spanish conquest. They preserve most of their old religion, incorporating it into Christianity, but the force of the old nature rules is reduced, and good land management survives largely because it pays. Fortunately, it is so well established and so grounded in everyday values that it survives. Moreover, "modern" industrial and monocropping regimes have always failed on the Yucatán Peninsula, because of soil and pest problems, and only Maya diversified agriculture succeeds (Anderson, 2005).

The Northwest Coast native peoples divide the world still differently. Humans are full participants in the world, linked closely by kinship and other social ties with animals, plants, and landscapes. Animal powers made the world what it is, and even mountains are sentient beings that once walked around and chose where to settle. Humans depend totally on other lives, and must treat them with care. The key word, according to a wide range of Northwest Coast thinkers and writers, is *respect* (Atleo, 2004; Brown & Brown, 2009; George, 2003). The English word respect undertranslates native terms, which imply reverence, concern, mutual support, personal closeness, and gratitude. The social sphere is the realm of everyday life as well as chiefly politics and war organization. The remote areas of nature, the true wildernesses, are spiritual: people seek mountaintops, remote lakes, isolated islands, and similar places for visions and spiritual discipline. An idea of wild behavior accompanies this; chiefs are controlled and beautifully dressed, while shamans, who are often close relatives of the chiefs, go about with rough clothing, wild unkempt locks, and strange behavior. Both chiefs and shamans get power from the wilderness, but shamans are closer to it.

The visions and spiritual experiences in the wild are the bases of life choices. Individuals receive spirit power, especially in vision quests around the time of puberty or adolescence. An individual seeks power from a guardian spirit, through isolation and self-mortification in remote and beautiful surroundings. The result may be ability in war, gambling, wood carving, or any other useful art. Healing usually requires a different and more rigorous discipline, with special healing spirits or powers obtained. (From a huge literature on this, an especially good indigenous account is Atleo (2004); classic "outsider" descriptions can be found in Benedict (1923) and Miller (1988, 1998, 1999). The Native American biologist Raymond Pierotti (2011; personal communication) points out that the wild thus is supremely important, but has none of the negative and frightening connotations that the word wilderness so often has in English.

Still a different conceptualization of wild and tame exists among the Uto-Aztecan peoples of Mexico (Hill, 1992). This contrasts a world of "flowers and songs"—which may be heaven or the wilderness—with the world of settlements. Highly elaborated by the Aztecs, this conception is found in a more straightforward form among the Yaqui, whose language is fairly close to the Nahuatl language of the Aztecs. The Yaqui divide the world into the everyday world of village communities, located in fertile river valleys, and the "flower world" (*sea ania*), the wilderness of remote deserts and brushlands (Evers & Molina, 1987; Painter & Spicer, 1986; Spicer, 1962). The village world is hot, dry, dusty, and thoroughly human- ized, with intensive agriculture, dense settlements, and everyday life and problems. The flower world is cool, moist, and abounding with life that is independent of humans. It is the realm of spiritual activities and hunting. It is also the realm of healing, partly because medicinal herbs come from it, and partly because people with healing personalities are attracted to it. Yaqui healers are supposed to be gentle, quiet, caring, and drawn into themselves, loving children and animals as well as the flower world. (This personality type is clearly identifiable as one that is rare in the general population, but extremely common among psychotherapists.) Exquisite songs emerge from contemplating the flower world. (Many can be found, with translations, in Evers & Molina, 1987. They are beautiful enough in translation, but I have had the pleasure of hearing Felipe Molina sing them in Yaqui, and in that form they are a supreme experience.) Hence, tame and wild do exist as separate named realms, but they depend on each other, and on each other's constant interpenetration and interaction.

Moving to another continent, the early Irish system of tame/wild op- positions is reflected in tales and epics (Anderson, 2010b). The surviving recensions hark back to pre-Christian or early Christian times, but are medieval in actual date, leaving us to infer what was really the state of belief in earlier days. Particularly interesting are two surviving works, the *Acallan na Senórach* (Tales of the Elders; Dooley & Roe, 1999) and *Buile Suibhne* or *Suibhne Geilt* (Mad Sweeney; Heaney, 1983; O'Keeffe, 1913), of uncertain date, although preserving apparently folkloric mate- rial dating to early Christian times. Both recount the coming of Chris- tianity to pagan Ireland. The former describes the meeting of the last mythical heroes of ancient Ireland with Saint Patrick (fifth century). The old heroes tell stories of their wild pagan days, and the book concludes with gratitude that things are peaceful and Christian now, but also with longing and sorrow for the lost heroic past. The latter is the story of King Sweeney, who resisted Christianity and was punished by "madness" until redeemed by Saint Moling (seventh century). The interest attaches to the dramatic contrast between the pre-Christian world and the world of Christianity. Pre-Christian Ireland is portrayed as a world of forests, wolves, feral cattle, and heroes locked in constant combat. Saint Patrick brought a world of settled, tamed life: monasteries, towns, proper church vestments and rituals, and orderly life. Actually, the contrast in these two tales is overdrawn for effect. Pre-Christian Ireland had long been fairly well tamed, and Irish Christianity from the start was close to pre-Christian Ireland in its attitudes toward nature. Christians were often hermits in the forest, living and communicating with animals.

The pre-Christian worldview portrayed in these texts was strikingly close to the Northwest Coast shamanic one. Sweeney's madness bears no resemblance to schizophrenia, depression, or other categories of mental illness; it is obviously a spiritual discipline close to shamanism. Mad Sweeney flies through the air, talks with animals and addresses trees, prophesies the future, sings powerful songs, and otherwise performs most of the acts considered diagnostic of shamanism (cf. Eliade, 1964). In these stories, the forests, moors, coasts, and hills are wild lands, where supposed madmen like Sweeney live with animals and plants. Even ordinary persons spend a great deal of time in the forests, hunting, gathering wild fruit or medicinal herbs, exploring, sending out war parties, and simply enjoying. The hero of the *Acallan na Senórach*, Finn McCool, prefers the belling of stags and singing of birds to the sounds of humanity. Sweeney prefers the howling of forest hounds to the talk of students in church.

This positive attitude toward nature and the wild was a general feature of Celtic literature in the medieval period, and for that matter of Celtic literature throughout time, including modern writings and folklore (Anderson, 2010b).

The Chinese language has a good translation of the word wild: *ye*. But *ye* refers to a desolate, lonely wasteland, which can be the result of battles and depopulation as easily as of nature taking its course. The Chinese word for wild or wilderness in the good sense is *shan shui*, "mountains and waters," which is also the general term for landscape paintings. Every viewer of Chinese art or reader of Chinese poetry knows that the Chinese idealize wild mountain-and-water scenes. (One Tang dynasty poet did equate ye with shan shui, but apparently for the surprise effect; Han-shan, 2000, p. 191.) The nearest equivalents to the word natural do not refer to nonhuman space but instead to what is inborn: *sheng*, "born," gave rise to *xing*, "innate, natural born"; there are many compounds of both words. *Ziran*, "spontaneous, self-generating," and related words make up a related cluster. They are equivalent to the term nature in the sense of "human nature."

The opposition of humans against the natural world was not unknown, and indeed was expressed in the basic triad of *Tian*, *di*, and *ren*, "heaven, earth, and humanity." But this very phrase expressed a needed unity transcending any of the three. The ideal was harmony with nature. The emperor had to officiate at sacrifices at the Temple of Heaven and Temple of Earth in Beijing. Harmony included doing things in the most simple or natural way possible. This harmony-in-simplicity was largely a religious ideal, not interfering with building canals and laying out cities, but it had a major effect on the landscape, preserving groves, marshes, and other useful features.

Chinese food production, until the mid-twentieth century, depended on a biological approach that necessitated designing with nature and maintaining a large percentage of the environment in a modified, yet not drastically modified, state. Agriculture depended on watershed management, frequently including forests. In marked contrast with the West, agriculture was extremely diversified, with hundreds of food crops and an agronomic regime dependent on integrating many crops into a single complex system (Anderson, 1988). Fishing, herb gathering, forestry, and other activities led to an awareness of the need to manage nature sustainably—a fact explicitly stated from very early times (Anderson, 2001; Tucker & Berthrong, 1998). The Communist "struggle against nature" came as a dramatic shift, resisted strongly at first, and China is now waking up to the disastrous effects of the change.

China in premodern times displayed a more mixed record in appreciation of the wild than did the societies described above. On the whole, Native Americans and early Celtic peoples saw the relatively wilder areas of the world as both good and important—realms of hunting, gathering, and spiritual activities. The Chinese saw this too, but were much more deeply agricultural and urban, and had distinctly ambiguous attitudes. Most people saw the wild simply as something to tame and farm, or use for purely extractive purposes, as a source of timber, game, and medicines. On the other hand, many viewed the shan shui as an arena for meditation, relaxation, escape, self-cultivation, and renewal, and this became a cultural ideal known to everyone through the ever-present art and songs. (Unlike the West, traditional China had no strong barrier between elite and folk culture. I have heard near-illiterate fisherfolk chant classical poems and quote classical philosophy, from oral tradition. Public temples and offices usually had some classical-style art on display, and ordinary people borrowed themes from this for weavings and folk arts.) This generally accompanied an explicit valuation of the natural in the sense of spontaneous and uninhibited. Sages sought to cultivate their inner nature (*xing*) by being out where natural processes go on unchecked.

By the third to fourth centuries BCE, this had hardened into an opposition between more human-centered philosophies and more nature-centered ones. At one extreme were the various schools that later came to be lumped together as "legalism." This term covers a number of ideas on statecraft, running from directions for managing bureaucracy to quite cynical apologies for autocracy. These notions state or imply that the non-human world is purely a source of material goods. In a more middle position were the Confucian scholars, who had a more positive attitude toward nature, and strongly advocated conservation and wise management (see Tucker & Berthrong, 1998), but generally saw nature as existing for humans to use. At the other extreme stood the Taoists (Girardot, Miller, & Liu, 2001), who saw humans as part of nature, and as needing to shed their artificial social and cultural traits, and merge into the vast flow of things. Within Taoism there were several subtraditions, ranging from a mild naturism and incipient natural science to the uncompromising transhumanism and transcendent mysticism of Zhuangzi (Graham, 1981). Later, Buddhism entered China, bringing a philosophy of valuing and being compassionate toward all beings. This inevitably fused with Taoist mysticism.

The result was all that much more dramatic for being counterposed to legalist and Confucian norms. Some scholars followed the tradition of Zhuangzi and Buddhist monasticism: they retreated from the "red dust"—a Buddhist term for the ordinary social world—to the mountains and waters. Many more kept their town jobs, but retreated to rural or remote environments for rest, renewal, meditation, poetic flights, and other pleasant occasions. (They were proverbially accused of being "Confucian at work, Buddhist in worship, and Taoist on vacation.") Either way, scholars, poets, and perfectly ordinary people—educated or not—intensely loved as well as valued flowers, trees, animals, mountains, waters, rural occupations, and above all, merging into the vast natural processes that comprise the Great Way (*tao*).

It is best to let the scholars speak for themselves. Chinese "poems" were actually chanted or sung, to traditional tunes, and were not for silent reading. When I did research in Hong Kong in the 1960s, many people, even illiterate fisherfolk who had learned them orally, could still do this. We must fall back on tuneless translations here, alas.

The popularizer, though not the inventor, of nature poetry was Tao Yuanming (367–421 BCE). His clearest statement of his mind-set was in a poem that has become well known in the English-speaking world through the translation by Arthur Waley (1946, p. 105):

I built my hut in a zone of human habitation;
Yet near me there sounds no noise of horse or coach.
Would you know how this is possible?
A heart that is distant creates a wilderness around it.
I pluck chrysanthemums under the eastern hedge,
Then gaze long at the distant summer hills.
The mountain air is fresh at the dusk of day;
The flying birds two by two return.
In these things there lies a deep meaning;
But when we would express it, words suddenly fail us.

Waley uses a rather explanatory style, but the Chinese is telegraphic. Classical Chinese was a semiartificial language developed from a sort of court speed writing, and poets made full use of the resulting brevity and ambiguity. Tao's last line literally says, "Want speak can't remember words." (The structure of five characters, grouped two and three, is typical of this style.) The ambiguities are fully deliberate. It could be the poet or all of us speaking; we may merely fail to find the right words for the scene, or we may have lost speech entirely, in a mystic state.

The great Buddhist hermit-poet Han Shan (eighth to ninth century) wrote:

Wise men, you have forsaken me;
Foolish men, I have forsaken you.
Being not foolish and also not wise
Henceforward I shall hear from you no more.
When night falls I sing to the bright moon,
At break of dawn I dance among the white clouds.
Would you have me with closed lips and folded hands
Sit up straight, waiting for my hair to go grey?

(cited in Waley, 1961)

Han Shan—"Cold Mountain"—is a peak in eastern China. We know nothing of the poet's real name—or even how many people Han Shan was, for the Han Shan material is in several styles and apparently accumulated over a hundred years. And apparently the alias became traditional. (One of the Han Shans was responsible for equating ye and shan shui above.)

Liu Tsung-yuan, another Buddhist-Taoist poet from around the eighth century, observed:

Thousand mountains no bird flies
Ten thousand paths no human track.
Solitary boat old man, split-bamboo raincloak,
Fishing, alone cold river snow.

The above verse is my translation this time, offering some sense of the Chinese. This poem is telegraphic and ambiguous even by classical Chinese standards, and I have tried to preserve that. The anonymous figure (the poet?) is invisible; we see only his raincoat. The doubled loneliness of "solitary" and "alone" is particularly intense in the Chinese text.

It is hard to imagine anyone getting more meaning into fewer words, but the Japanese carried the Chinese tradition to at least equal heights, and their haiku have only seventeen syllables (Liu's Chinese has twenty). Here is a haiku by Issa (1763–1828), in Japanese, followed by my translation (based on MacKenzie, 1957, p. 2):

Akikaze ya!
Hyoro-hyoro kama
no kageboshi

The autumn wind!
It shakes
the mountain's shadow

The effect of moving air masses, making the shadow seem to tremble, becomes a comment on power and transiency.

In addition to the Japanese, the Koreans and Vietnamese picked up Chinese nature poetry and nature ideology, and developed the philosophy in their own ways.

Obviously every poem has its own message, but the grounding philosophy is fairly uniform. We have to escape the red dust to see things clearly. When we do, we can appreciate natural beauty, see how our own lives should be lived, and recognize the parallels between outward nature and inward nature—the personhood lying below or behind the self. We can also appreciate the lives of ordinary rural people: woodcutters, fishers, and farmers. The poets range, however, from complete mystics who utterly lost themselves in the Tao to self-conscious, forward-looking Confucians who loved trees, flowers, and birds, yet loved their families, friends, and professional work more.

The philosophies in question are reflected in landscape art, garden design, architecture, temple precincts, and even music as well as literature.

Chinese landscape paintings usually show a few tiny figures contemplating the vast scenery that dwarfs them into near insignificance. Sometimes the figures are emphasized more, but never is the landscape a mere backdrop for a portrait, as it is in much of European art. (Take Leonardo da Vinci's *Mona Lisa*, for example; many viewers do not even realize that she is backed by a detailed and beautiful view of her northern Italian homeland.)

Sustainable Relationships

These vignettes are only a sampling of the representations of nature found in various worldviews. One thinks of the soaring beauty of Australian aboriginal art, compassion for all beings of Buddhist philosophy, deep nature philosophy of northern South American peoples (Reichel-Dolmatoff, 1971, 1976, 1996), and thousands of other cultural worldviews.

Worldviews express real-world relationships, with appropriate allowance for metaphorization, idealism, and nonsynchronized changes, and also express the morality that guides environmental management. The Maya worldview both mirrors and guides an overall land management strategy. Northwest Coast views reflect and inform a nonagricultural as well as nonintrusive, but intensive and pervasive, way of using the environment; resources are left in a minimally transformed state, yet carefully managed for sustained yield. The Chinese traditional system involved leaving wildlands relatively undeveloped, but systematically cropped for sustainable yields of natural resources; lowlands, however, were intensively farmed, although again with sustainable management. Urbanization led to large areas being transformed profoundly, with gardens and parks as the only relief; nevertheless, these brought nature—miniaturized—into the city.

We have too few descriptions of these other ways of knowing and caring. Ethnographers rarely stay long enough to explore the depths of the cultures they study, and more and more ethnographers today are concerned only with "victims"—people reduced to helpless, voiceless pawns caught in "globalization" and "neoliberalism"; we need to listen more to local voices (Escobar, 2008). And even the best ethnography is inadequate if there is only one. It has taken us 150 years to scratch the surface of Australian aboriginal worldviews. Centuries of intensive scholarship on Chinese poetry still leave much to be said.

We need detailed studies over time and from many points of view. We have these for some Native North American peoples (such as the Navaho and Hopi) and a few other traditional peoples worldwide.[2] One case deserving of attention is that of the Winnebago of the northern Midwest; their environmental views and strategies were the subject of classic, pathbreaking ethnography by Paul Radin (1916), with reassessment and restudy by Thomas Davis (2000). The closely related Menomini have been studied by a team involving cognitive psychologists and anthropologists (Medin, Ross, & Cox, 2006). These and several other works give us a solid foundation. Such cases are rare; even majority European and US societies are poorly studied in this regard. Literary criticism of famous books is no substitute for ethnography.

Most early or traditional societies recognize that people depend directly on the rest of the world, and thus see "people in nature" rather than "people versus nature." The latter may have begun in Mesopotamia, but took full form in the Roman Empire. The Romans developed huge slave-managed estates. The medieval period saw these grow into true plantations. The Age of Enlightenment was ironically the peak period of the slave trade, which obtained labor for ever-larger plantations. This has led some scholars to find a reaction against slavery in Enlightenment thought, and others to find an expression of the hierarchical, ordered view of the universe that slave plantations represented. In fact, one can find both (cf. Foucault, 1966/1970; Pierotti, 2011). The substitution of machines for slaves is a recent phenomenon, forced by the elimination of slavery (on religious and moral grounds) in the nineteenth century.

The world is now based on the agricultural economy developed in Mesopotamia. It eliminates all natural features to grow endless expanses of monocrops, now often genetically engineered and genetically identical. California's rice farmers use laser-guided machinery to level their paddies to a tolerance of a fraction of an inch, sow by airplane, drop pesticides by airplane, and harvest with large combines; the difference from Chinese agriculture is obvious. One expects few poems from the California rice landscapes. If this really increased production it might be a good idea, but in fact Chinese labor-intensive agriculture can produce more rice per acre, per energy unit, and per seed (machines just aren't perfect). And of course modern nature-destroying agriculture and urbanization are not sustainable in the long run. Experience, from US organic farms to Southeast Asian dooryard gardens (Anderson, 1993a, 1993b), shows that this extensive-monocrop regime is the result of cultural bias, not economic or ecological superiority.

This machine-and-monocrop world, however, comes from only one strand of Western thought on the environment. The ancient Greeks and Romans already had countertraditions, idealizing small free farming and the natural environment. From early medieval times, the Celtic input to European thought has been enormous, as has input from other, more environmentally conscious parts of Europe, from Scandinavia to Russia. Landscape art, Romantic period poetry, the philosophy of Henry David Thoreau, the science of John James Audubon and George Perkins Marsh (1864/2003), gardening, urban planning, and countless other Western creations show the continuance of respect for nature in Western traditions. The radically different attitudes reflected in the first two books of Genesis were reinforced by the radically different attitudes coming to European thought from the Roman core and Celtic-Germanic-Slavic fringe. It is with us today in the contrast of agribusiness and urban sprawl with national parks and wildernesses as well as other reserves with varying degrees of management.

If we are to return to a concept of people in nature, we must confront the differences between the various worldviews. The East Asian (and countless other) cases remind us that societies can manage unevenly, being good managers in some areas and poor in others. And not all traditional societies practice conservation or long-term resource management. Some do not even have the concept (Beckerman, Valentine, & Eller, 2002; Hames, 2007). Some simply use anything they can find, as soon as they find it. This is especially true if they are Westernized societies, or if they are small and technologically simple groups that live in exceedingly low population densities in rich environments, plus have little impact on their overall systems (Beckerman, Valentine, & Eller, 2002). In addition, many groups have abandoned sustainable management during periods of war and violent unrest, for obvious reasons. Still others manage well, but intensively, so that the environment is nothing like a natural one; this is true of the wet-rice systems of lowland Southeast Asia. Even sparsely populated hunter-gathering peoples can considerably transform and manage the wild (Kay & Simmons, 2002). The idea of the ecologically noble savage must be replaced by the recognition that there are thousands of traditional cultures, each with their own management rules and broader cosmovision that culturally constructs those rules in religious or ideological frameworks.

The split-personality situation of Western civilization is probably more typical than is smooth consistency. Significantly, the studies of the Winnebago and Menomini turned up profound differences in worldview among different individuals, even within these tiny cultural groups (Radin, 1916, 1927, 1937/1957; Medin et al., 2006). Culture is not a straitjacket that makes all its bearers think alike. Moreover, all cultures constantly borrow and blend, and people draw on many repertoires, cultural and other; the idea of closed, monolithic, distinct cultures is incorrect.

Culture is actually a repository of ideas and representations, which then serve as sources of guidance for anyone influenced by the culture, or cultures, in question. Inevitably these ideas and representations convey diverse messages, which can be interpreted in even more diverse ways. Then the task of the culture bearers is to get people to be at least consistent enough to allow society to function. Some messages remain ambiguous; the tension between dominion and stewardship is still with us. Others are sharply enforced; the conserving behaviors involved in Northwest Coast "respect" were enforced quite strictly, in the full knowledge that failure to enforce them almost guaranteed starvation when prey populations collapsed.

Managing often involves deferred gratification or outright sacrifice of one's own interests to those of the group. People do not always do such things willingly. Evidence suggests that they are actually less willing to sacrifice today for much larger future personal benefits than to sacrifice for immediate benefits to others around them (especially family and friends, of course). A great deal (one might even say all) of social morality is devoted to these two ends. Resource conservation morality provides a major theme in the mythology of the Northwest Coast peoples; all groups apparently shared stories of game and fish depletion in the past by careless, disrespectful people, who overdrew resources. Chinese stories about those who violated feng shui rules and other resource-related religious taboos are universal; I have heard many. The Celtic literatures evidently had considerable environmental teaching function. Sweeney sings a song of the trees that appears related to a long-lasting folk poem teaching children what woods to use for firewood. This poem was in evidence in medieval Wales and survived (in English) in Wales up into the twentieth century (Yeldham, 2007). Caring for and about nature on a wider, deeper level certainly emerges from countless medieval saints' lives, folk charms (Carmichael, 1899/1992), and other folk literature.

In short, the representations of nature in traditional literature are not there for ornament. They are there to build concern and teach wise use. They create bonds with the nonhuman world.

This is most obvious in the many cases worldwide in which human kinship systems are directly extended to include nonhuman lives. This sort of "totemism" (in the widest sense) has been analyzed by anthropologists interested in what it tells us about human kinship (Durkheim, 1912/1995; Lévi-Strauss, 1963). Many anthropologists have also made the point that in a concrete sense, including animals and plants in human kinship groupings creates a brotherhood and sisterhood, a family relationship, in which people will care for their "other-than-human" relatives. This is merely an extension of the near-universal religious practice of referring to fellow communicants as brothers and sisters, or brethren.

Common Themes

Any type of involvement, good or bad, produces emotion; the more complex and deep the involvement, the more complex and deep the emotion. Karl Marx knew this, in spite of his materialism. People are most involved, first, in their families and secondarily in their work (cf. Frankl, 1959). This is the key to traditional religion and environment use: total use of the environment and continual interaction with it produces a deep, rich personal experience. This involves both cognitive learning and emotional deepening. It is then culturally constructed as an ideological representation system—that is, as religion or a functional equivalent.

As mentioned earlier, Taylor (2010) uses the term dark green religion for the intense, loving experience of and devotion to nature, whether religiously expressed or not. He traces it in the United States from Thoreau through John Muir and Aldo Leopold, to the incredibly diverse modern activists united only by desire to conserve the wild and other than human. Taylor admits that this stretches the usual definition of religion, but the word is so poorly defined in any case (see also Atran, 2002) that he can define deep green religion and stay within reasonable bounds. We do, however, need to analyze these matters further.

Living with nature and managing to preserve biodiversity need not be religiously represented, though religion certainly helps. In Southeast Asia, small-scale societies ("hill tribes") along with traditional Hindu and Buddhist states have managed the natural landscape quite well, and have religious and cultural rules as well as ideals that reinforce and sanction the management strategies. Much of the region, however, converted to Islam—or locally, Christianity—centuries ago, and continued to manage as well as the other societies did, in spite of the loss of the old religious guidelines. Conservation here was based on the commonsense need to manage whole landscapes. The traditional wet-rice agriculture could use less than 10 percent of the land; other forms of agriculture required forest fallow or forest creation; and people had learned to live by drawing on all classes of resources and all landscape types. Modern Western-style cultivation has come to the region with disastrous effect, importing mono-crop agriculture that is poorly suited to the land or sociocultural systems.

What to Do Next

To make a long story short, religion, secular cultural representations, economic necessity, or simple common sense can all guide land management. Religion has the enormous advantage that it can deploy real fear: the waster or rule breaker will be punished with disease, death, or hellfire. But secular rules enforced by guilty conscience can be effective. Rational self-interest is a problem rather than a useful mechanism, since the cases where enforcement is needed are precisely those where rational self-interest goes against social welfare. (What follows is developed further in my book *The Pursuit of Ecotopia* [Anderson, 2010a].)

At present, it would seem that almost everything in the world is going in the wrong direction. Monocrop agriculture, urbanization, reduction in biodiversity, and pollution are devastating the environment. The idea of people in nature is hard to establish when people are in fact living in cities where even a single tree is an increasingly rare sight. Perhaps worst of all, though, is the loss of the old religions, ideologies, and cosmologies that taught people to care for the world as well as sacrifice for the common good. The rising ideologies today are authoritarian militarism and the narrow form of "rational self-interest" that considers only immediate financial returns. These lead to displacement of small indigenous societies, or conversion of their members to the "modern" ways of thinking.

A particularly revealing and ironic effect is what we might call "conservation against itself." In many areas, traditional people have been forced off their lands in the name of conservation. I have seen this happen with the Maya on the Yucatán Peninsula (Anderson, 2005). There is an excellent review of the worldwide problem by Paige West, James Igoe, and Dan Brockington (2006), although the situation has gotten worse since their article (Igoe, 2010). It appears that the usual offenders are either large government agencies or conservation nongovernmental organizations that depend heavily on large corporate sponsors for funding, and not infrequently, the corporations have some financial interest in getting indigenous people off their lands.

The problem is much more than the loss of livelihood for displaced individuals. It involves the loss of the entire land management systems the cultures developed over millennia. Ironically, it is precisely where land management was best that the forests, wildlife, and wetlands are in the best shape—and therefore it is often the best managers that are the ones ruled off the land, since their traditional homes are the most valued for biodiversity conservation. This is certainly the case with the Maya, and with the Maasai studied by Igoe, among many others. The land suffers for lack of good management; the people often lose their good management strategies without their land.

In a few areas, this is being reversed, possibly too late, but at least with good intentions. The Timbisha Shoshone are managing lands in Death Valley once more. The Australian Aborigines have gone back to managing their more remote areas, with varying degrees of cooperation (or sometimes, foot-dragging) by Anglo-Australian stakeholders. Many South American tribes have management rights over rain forests that until recently were up for grabs.

Dealing with these problems must obviously be complex. The security of a title and management rights for traditional small-scale societies is clearly one place to start. Fixing the global economy so that the costs of production are actually specified by the producers is another necessary starting point. We cannot continue to allow those who convert tropical forests to monocrops to keep passing off the real costs on the local poor. We cannot continue to subsidize oil companies that displace indigenous peoples and destroy their lands.

Obviously, though, the major need is cultural representation systems that will make people care and be responsible.

Simple experience with nature, the wild, or even ordinary tame plants and animals is the clear place to begin, but it is never enough. Most people care about nature somewhat. Yet they do not prioritize it; nor do they have a sense of people in nature (as opposed to people versus nature). We also lack effective mechanisms for motivating responsibility. This is where traditional societies had an advantage. Today, ordinary people still mess up the environment, and lawmakers still pass laws on the assumption that conservation is a dispensable cost, not an indispensable necessity. This is bad enough in the case of biodiversity, but it becomes truly insane when water is the resource at issue; it is amazing to find almost complete irresponsibility about water conservation in places as desperately short of that commodity as China (Pomeranz, 2009) and Arizona (Anderson, 2010c).

The conservation organizations that seem to be doing the best job are those that use a multipronged approach. They have beautiful literature for the general public, conduct independent research, and engage actively in politics. Political engagement for a cause appears to be not only necessary for itself but also a partial equivalent to the religious taboos of old; it gets people involved socially, engages their consciences, and motivates them to act for the common good. At least some of the most successful not only do all the above but have direct tangible benefits as well. They save ducks for duck hunting (Ducks Unlimited), manage sanctuaries as well as run camps and workshops (Audubon Society), or do serious research and development (California Native Plants Society). Conversely, they do *not* have extremist programs or take radical political positions. They also are not dependent on large corporate sponsors (although not automatically ruling them out).

The finding that *what matters are pragmatic, on-the-ground ideas about management* is particularly important. Actual economic facts are generally less regarded; people manage the way they know how, which is usually the way their culture has done it for thousands of years. On the other hand, excessively abstract, philosophical, or literary ideas do not always have much relationship to what people actually do. Japanese Buddhist ideals do not stop the Japanese from fishing whales and tuna to extinction. Extreme philosophical principles may provoke a counter-reaction. The West may have turned against nature, but that very fact made many Westerners argue strongly *for* nature and conservation. (Farther afield, consider how little the claim that "Americans are individualistic" tells you about the play of conformity, imitation, faddism, and occasional real originality that US life really entails.) What matter in the actual interaction with nature are the beliefs that directly guide such immediate interaction.

Such goals as diversifying agriculture and conserving wildlands depend on our developing a more holistic worldview. The idea is not to disregard economics but instead to recognize that small mixed farms, efficient use, broad environment care, and collective planning for resource management are economically better than destroying everything natural for extensive monocrop agriculture.

This means that education has to change drastically.[3] Cramming students with random facts to memorize for multiple-choice tests is not optimal for environmental education. In fact, we have got things exactly the wrong way around. Students have to get out to the land—to wildlands, farms, ranches, parks, and everything (Louv, 2005). They have to *do* things: count, collect, experiment, raise plants, and feed animals. They have to get a sense of where our food and fiber come from. In the classroom, they need to think, analyze, write reports, write stories, write poems, or write anything at all. Texting and Facebook posting prove that students today have a real and felt need to write. They even create stories and poems. They are doing this for pleasure, without even knowing it was once part of education. An education consisting entirely of mindlessly filling in bubbles on a form teaches children to hate the whole education process, and they often wind up antiscience and anti-intellectual, giving us the voters of today who reject public education, evolution, global warming, and environmental concerns in general.

Experience with traditional cultures—not only indigenous peoples, but Anglo-American rural culture too—suggests that another valuable way to teach is through stories, especially stories about the immediate area. Every traditional culture about which I know anything at all uses this approach. Myths, tales, personal stories, and other narratives are second only to actual guided practice as teaching vehicles (Cruikshank, 1998, 2005).[4]

These are the methods used by traditional societies to enculturate the young in the immediate ecological and economic worldviews that matter to them. The methods work because they are precisely the ones that teach the kinds of views that I have been describing: direct, on-the-ground principles for working and acting. Unfortunately, our educational system has been drifting further and further away from such views along with the methods that best teach them. We produce philosophers who can talk about environmentalism in the abstract, but do not deal with real economic choices; we produce engineers and developers who can build dams and drain marshes, yet do not examine the principles on which such plans are based. We certainly need abstract thinkers and factually grounded workers, but we particularly need thoughtful workers.

We should follow traditional societies in couching conservation in aesthetic, moral, and personal terms as well as philosophical, utilitarian, and ecological ones. The traditional systems that interest me all use powerful visual, verbal, and performance art to establish as well as motivate concern for nature. This seems, from simple comparative ethnology, to be absolutely necessary.

Conclusion

Probably the most significant point I have learned from traditional societies is that caring for nature is a part of caring for people. Responsibility for the environment is part of responsibility in general.

This has several implications. First, we cannot get people to care for the environment by talking about the environment as if it were an alien, different thing, outside daily life. The language of citizenship and personal responsibility is more valuable. Too often, people feel involved in and dependent on the economy, but feel the environment is a distant and irrelevant concern; this produces the wildly misleading "jobs versus owls" rhetoric. I was rather appalled to find even Nobel Prize winner Amartya Sen (2009, p. 227) writing, "We could choose to use our freedom to enhance many objectives that are not part of our own lives in a narrow sense (for example, the preservation of animal species that are threatened with extinction)." Sen, of all people, should know that our lives, even "in a narrow sense," are economically, emotionally, and morally bonded inseparably and tightly with other animals' lives. Some of those protesting the displacement of indigenous peoples have also contended that conservation is an irrelevant concern, clashing with economics. The indigenous peoples themselves would never argue that way. In reality, the opposition is between saving people in nature and losing both the people and the environment. We should phrase our discourse accordingly.

This means that we have to recognize that individual trees, birds, and fish have at least *some* moral standing as persons. Perhaps we cannot give full humanity to other animals, and certainly we need feel little kinship or mercy toward malaria trypanosomes or tapeworms. But the idea that an entire ecosystem should be sacrificed merely on the whim of one rich person, or merely for the transient comfort of a few, seems to me not morally acceptable.

Perhaps we can bring the world's religions and ideologies together around the common themes of respecting life and valuing living beings (cf. Atleo, 2004). Above all, we need to place the highest possible value on continued learning and the progress of knowledge, emotional and spiritual as well as pragmatic and factual.

Author Note

I would like to acknowledge Jolina Ruckert and Peter Kahn for their valuable editorial help.

Notes

1. *Natura* is cognate with *nascitur*, "to be born," and originally referred to the womb and/or inborn qualities.
2. Some idea of what we have lost is conveyed by the fact that the relatively small Navaho tribe has provided thousands, and perhaps tens of thousands, of pages

of superb religious poetry (e.g., Luckert, 1979; Matthews, 1887, 1902; Zolbrod, 1984), almost all concerning nature in some way. Similarly, the ancient Jews, largely during their Babylonian captivity, wrote down their traditional canonical literature, giving us the Hebrew Bible. From what little research has been done, it is fairly evident that a large percentage of the world's cultures once had similarly voluminous, beautiful, and instructive oral traditions. But we have lost almost all of them, and the time is gone when they can be recorded; most are extinct or nearly so. Recent ethnographies rarely provide any materials of this nature.

3. For materials and references, see my Web site, www.krazykioti.com.

4. Correcting a wayward child is often done indirectly: an elder will reminisce about the foolish things they did in their youth and how they learned better. The foolish things are, of course, the things that the wayward child is doing right now. It is a way of reprimanding a child without making the child lose face, or feel isolated and uniquely guilty. This was a major feature of my own childhood, and it is widely described around the world.

References

Anderson, E. N. (1988). *The food of China*. New Haven, CT: Yale University Press.

Anderson, E. N. (1993a). Gardens in tropical America and tropical Asia. *Biotica n.e.*, *1*, 81–102.

Anderson, E. N. (1993b). Southeast Asian gardens: Nutrition, cash, and ethnicity. *Biotica n.e.*, *1*, 1–12.

Anderson, E. N. (2001). Flowering apricot: Environmental practice, folk religion, and Daoism. In N. J. Girardot, J. Miller, & L. Xiaogan (Eds.), *Daoism and ecology: Ways within a cosmic landscape*. Cambridge, MA: Harvard University Press.

Anderson, E. N. (2005). *Political ecology of a Yucatec Maya community*. Tucson, AZ: University of Arizona Press.

Anderson, E. N. (2010a). *The pursuit of ecotopia*. Santa Barbara, CA: Praeger.

Anderson, E. N. (2010b, June 13). Seeing the natural world in medieval Ireland. *Krazy Kioti—The Gene Anderson Webpage*. Retrieved from http://www.krazykioti.com/articles/seeing-the-natural-world-in-medieval-ireland.

Anderson, E. N. (2010c, July 6). Water. *Krazy Kioti—The Gene Anderson Webpage*. Retrieved from http://www.krazykioti.com/articles/water/.

Atleo, E. R. (2004). *Tsawalk: A Nuu-chah-nulth worldview*. Vancouver, Canada: University of British Columbia Press.

Atran, S. (2002). *In gods we trust*. New York, NY: Oxford University Press.

Bartra, R. (1994). *Wild men in the looking glass*. Ann Arbor, MI: University of Michigan Press.

Bartra, R. (1997). *The artificial savage*. Ann Arbor, MI: University of Michigan Press.

Beckerman, S., Valentine, P., & Eller, E. (2002). Conservation and native Amazonians: Why some do and some don't. *Antropologica, 96,* 31–51.

Benedict, R. F. (1923). *The concept of the guardian spirit in North America, memoir 29.* Menasha, WI: American Anthropological Association.

Brown, F., & Brown, Y. K. (2009). *Staying the course, staying alive* (CD and folder). Vancouver, BC: Biodiversity BC.

Carmichael, A. (1992). *Carmina gadelica.* Hudson, NY: Lindisfarne Press. (Original work published 1899).

Cruikshank, J. (1998). *The social life of stories: Narrative and knowledge in the Yukon territory.* Lincoln, NE: University of Nebraska Press.

Cruikshank, J. (2005). *Do glaciers listen? Local knowledge, colonial encounters, and social imagination.* Vancouver, Canada: University of British Columbia Press.

Davis, T. (2000). *Sustaining the forest, the people, and the spirit.* Albany, NY: State University of New York Press.

Dooley, A., & Roe, H. (1999). *Tales of the elders of Ireland: A new translation of Acallam na Senórach.* Oxford, UK: Oxford University Press.

Durkheim, E. (1995). *The elementary forms of religious life.* K. E. Fields, (Trans.). New York, NY: Free Press. (Original work published 1912).

Eliade, M. (1964). *Shamanism: Archaic techniques of ecstasy.* London, UK: Routledge and Kegan Paul.

Escobar, A. (2008). *Territories of difference: Place, movement, life,* Redes. Durham: Duke University Press.

Evers, L., & Molina, F. S. (1987). *Yaqui deer songs: Maso Bwikam. A Native American poetry.* Tucson, AZ: University of Arizona Press.

Foucault, M. (1970). *The order of things: An archaeology of the human sciences* [Les mots et les choses]. New York, NY: Random House. (Original work published 1966).

Frankl, V. (1959). *Man's search for meaning: An introduction to logotherapy.* Boston, MA: Beacon.

George, E. M. (2003). *Living on the edge: Nuu-chah-nulth history from an Ahousaht chief's perspective.* Winlaw, Canada: Sono Nis Press.

Girardot, N. J., Miller, J., & Liu, X. (Eds.). (2001). *Daoism and ecology: Ways within a Cosmic Landscape.* Cambridge, MA: Harvard University Press.

Graham, A. C. (1981). *Chuang Tzu: The inner chapters.* London, UK: George Allen and Unwin.

Hames, R. (2007). The ecologically noble savage debate. *Annual Review of Anthropology, 36,* 177–190.

Han-shan. (2000). *The collected poems of cold mountain.* (R. Pine [B. Porter], Trans.). Port Townsend, WA: Copper Canyon Press.

Heaney, S. (1983). *Sweeney astray.* London, UK: Faber and Faber.

Hessel, D., & Reuther, R. R. (2000). *Christianity and ecology: Seeking the wellbeing of Earth and humans.* Cambridge, MA: Harvard University Press.

Hill, J. (1992). The flower world of old Uto-Aztecan. *Journal of Anthropological Research*, 48, 117–143.

Igoe, J. (2010). The spectacle of nature in the global economy of appearances: Anthropological engagements with the spectacular mediations of transnational conservation. *Critique of Anthropology*, 30(40), 375–397.

Jackson, K. H. (1971). *A Celtic miscellany*. London, UK: Penguin.

Kay, C. E., & Simmons, R. T. (Eds.). (2002). *Wilderness and political ecology: Aboriginal influences and the original state of nature*. Salt Lake City, UT: University of Utah Press.

Kovacs, M. (1985). *The epic of Gilgamesh*. Stanford, CA: Stanford University Press.

Lévi-Strauss, C. (1963). *Totemism*. R. Needham, (Trans.). Boston, MA: Beacon.

Lévi-Strauss, C. (1964). *Le cru et le cuit*. Paris, France: Plon.

Louv, R. (2005). *Last child in the woods*. Chapel Hill, NC: Algonquin Books of Chapel Hill.

Luckert, K. W. (1979). *Coyoteway, a Navaho holyway healing ceremony*. Tucson, AZ: University of Arizona Press.

MacKenzie, L. (1957). *The autumn wind: A selection from the poems of Issa*. London, UK: John Murray.

Marsh, G. P. (2003). *Man and nature*. D. Lowenthal, (Ed.). Seattle, WA: University of Washington Press. (Original work published 1864).

Matthews, W. (1887). The mountain chant. *Washington, DC: Bureau of American Ethnology, Annual Report*, 379–467.

Matthews, W. (1902). The night chant. *American Museum of Natural History, Memoir VI*, 1–332.

Medin, D., Ross, N. O., & Cox, D. G. (2006). *Culture and resource conflict: Why meanings matter*. New York, NY: Russell Sage Foundation.

Miller, J. (1988). *Shamanic odyssey: The Lushootseed Salish journey to the land of the dead in terms of death, potency, and cooperating shamans in North America*. Menlo Park, CA: Ballena Press.

Miller, J. (1998). Middle Columbia River Salishans. In D. Walker (Ed.), *Plateau handbook of North American indians* (Vol. 12). Washington, DC: Smithsonian Institution.

Miller, J. (1999). *Lushootseed culture and the shamanic odyssey: An anchored radiance*. Lincoln, NE: University of Nebraska Press.

O'Keeffe, J. G. (1913). *Buile Suibne (The frenzy of Suibne), being the adventures of Suibne Geilt, a middle-Irish romance*. London, UK: Irish Texts Society.

Painter, M. T., & Spicer, E. (1986). *With good heart*. Tucson, AZ: University of Arizona Press.

Pierotti, R. (2011). *Indigenous knowledge, ecology, and evolutionary biology*. New York, NY: Routledge.

Pomeranz, K. (2009). The great Himalayan watershed: Agrarian crisis, megadams, and the environment. *New Left Review.* Retrieved from http://www.newleftreview.org/A2788.

Radin, P. (1916). The Winnebago tribe. *Washington, DC: Bureau of American Ethnology. Annual Report, 37,* 33–550.

Radin, P. (1927). *Primitive man as philosopher.* New York, NY: Appleton.

Radin, P.(1957). *Primitive religion.* New York, NY: Dover. (Original work published 1937).

Reichel-Dolmatoff, G. (1971). *Amazonian cosmos: The sexual and religious symbolism of the Tukano Indians.* Chicago, IL: University of Chicago Press.

Reichel-Dolmatoff, G. (1976). Cosmology as ecological analysis: A view from the rain forest. *Man, 11,* 307–316.

Reichel-Dolmatoff, G. (1996). *The forest within: The world-view of the Tukano Amazonian Indians.* London, UK: Themis-Green Books.

Sen, A. (2009). *The pursuit of justice.* Cambridge, MA: Harvard University Press.

Spicer, E. H. (1962). *Cycles of conquest.* Tucson, AZ: University of Arizona Press.

Taylor, B. (2010). *Dark green religion: Nature spirituality and the planetary future.* Berkeley, CA: University of California Press.

Tirosh-Samuelson, H. (Ed.). (2002). *Judaism and ecology: Created world and revealed word.* Cambridge, MA: Harvard University Press.

Tucker, M. E., & Berthrong, J. (1998). *Confucianism and ecology: The interrelation of heaven, earth, and humans.* Cambridge, MA: Harvard University Press.

Waley, A. (1946). *Chinese poems.* London, UK: George Allen and Unwin.

Waley, A. (1961). *Chinese poems* (2nd ed.). London, UK: George Allen and Unwin.

West, P., Igoe, J., & Brockington, D. (2006). Parks and peoples: The social impact of protected areas. *Annual Review of Anthropology, 35,* 251–277.

Yeldham, C. (2007). Living history: Cooking and fire. *Petits Propos Culinaires, 84,* 121–126.

Zolbrod, P. (1984). *Dine' Bahane': The Navaho creation story.* Albuquerque, NM: University of New Mexico Press.

9

Five Feathers for the Cannot Club

Dave Foreman

Part I

If we look far and deep and wide, the key question that looms for Man is, "How do we fit in and live with all the other Earthlings for the long haul?"[1] Now I know it is unlike Man to look or think long and wide and deep. We seem hemmed in by short, narrow, shallow sight. By thinking with such blinders, we see ourselves as the only thing that has meaning and our few years as all time.

One hundred and fifty years ago, Charles Darwin lengthened, widened, and deepened our ken. His "theory" of evolution, though, is truly a handful of deep insights or "theories" that overthrew most philosophy before his time (Mayr, 2001). Darwin's bedrock insight is that all life comes from one forebear—the last common ancestor, or as biologists call it today, LCA. Thus, we are all kin, from microscopic wrigglers to cloud-catching coastal redwoods and burly great blue whales. Such a kinship, where we can call any other Earthly being "cousin," should broaden our view of life.

Moreover, Darwin took the tale of life back far, far longer than the six-thousand-some years given by Genesis. The oldest fossils we have unearthed peg life as being at least three-and-a-half-billion years old. Were we truly sapiens or even half as wise as we think we are, we would have greatly lengthened our view of life by now.

Further deepening life's flow, Darwin saw that evolution has not worked with goals in mind, nor has it been overseen or led in any way. Paleontologists, such as the late Stephen Jay Gould (1991), chide our high and mighty gall with the sharp understanding that therefore, Man is not the unerring outcome or end point of hundreds of millions of years of "life's descent with modification" but rather a happy or unhappy (hinging on what kind of earthling you are) happenstance. Belying Gandalf along with other wizards and sages, we were not "meant to be." Nor is anything Man has done in its flicker of time been meant to be. We happened to become, just as did deep-sea fish with gleaming flesh lanterns hanging in front of their nightmare mouths.

We only happened to be.

This may be the hardest, most frightening teaching from evolutionary biology and paleontology. It might well be some of why most *Homo sapiens* do not believe biology has much to do with us. That we were not meant to be but instead only happened to be is likely the most revolutionary idea in Man's tale, putting Thomas Jefferson and Karl Marx in the shade. The worrisome upshot of happenstance for many is that if we are not thoughtfully made by God—and given stern commandments that carry dreadful punishment for not following them—what is there to spur good behavior, what grounds for weighing anything as good or bad? Forsooth, though, of the many folks I've known in my three score and four on both sides of the believer/freethinker cleavage, it has been the freethinkers on the whole who have been more likely to hew to the golden rule, to "treat your neighbor as you would be treated," and be righteous and upstanding. Nonetheless, this fear (of how the happenstance of Man's becoming undercuts any spur to be good) gives me one of my key jobs: to see if I can crack this hard-shelled nut and find the sweet nutmeat of goodness within. In truth, others have already done so. Aldo Leopold (1953, p. 146), a great conservationist, ecologist, and philosopher of the first half of the twentieth century, did away with any worry about whether the nut held something good when he wrote, "The last word in ignorance is the man who says of an animal or plant: 'What good is it?'"

Building on the kinship-of-all-Earthlings bedrock, ecologists see living beings along with geologic doings and weather as living as well as working in ecosystems or "communities." Leopold (1949, p. 204) wrote the word "land" to mean the community of "soils, waters, plants, and animals." He further wrote under the heading "The Community Concept" that "all ethics so far evolved rest upon a single premise: that the individual is a member of a community of interdependent parts" (ibid., p. 203). From that, he brings us the upshot: "In short, a land ethic changes the role of *Homo sapiens* from conqueror of the land-community to plain member and citizen of it. It implies respect for his fellow-members, and also respect for the community as such" (ibid., p. 204).

So maybe Leopold, writing in 1948 or 1949, got us somewhere near to answering the question with which I began: "How do we fit in and live with all the other Earthlings for the long haul?" In all of Earth's communities—or *neighborhoods*, as I'd rather call them, for the cozier feeling—we need to be a plain member and citizen; in other words, a good neighbor, one who does not lord over the others. Perhaps being a good neighbor—whether we are a special creation or an only-happened-to-be, fleeting outcome of evolution—is the root for behaving in a good way (*ethically*). Leopold and others have thought that the grounds for ethics are about living together in a community or neighborhood. Indeed, how are quarrelsome, nasty ground apes such as we to get along with each other? Archaeologists and anthropologists have lately shown how warlike (with other bands) and murderous (within our own bands) we have been for many long years (Keeley, 1996; LeBlanc & Register, 2001 Stanford, 1999). Love-your-neighbor guidelines make our living together more workable. I underline the thought of all life living in neighborhoods since we *Homo sapiens* most of all need a lodestar of behavior to live in neighborhoods. Not seeing other Earthlings as our neighbors lets us deal with them as being without worth. This is why Leopold beseeched us to be plain members and citizens of the land community—so that we could live as friendly neighbors with all life, much as folks he had grown up with in Burlington, Iowa, and as folks did in his university town of Madison, Wisconsin. Small-town America has always seemed more neighborly than elsewhere.

Now, given the bedrock and the flooring on it we have built, where do we begin?

Here.

Life is good.

Manyfold, tangled life is better.

Manyfold, tangled life not hobbled by Man's will is best.

What do I mean? And how can I look on life and call it good if it all comes from happenstance?

By "life is good," I am not writing a television commercial about sitting with your buddies in front of a wide-screen television for a Superbowl party with Budweiser, while wives and girlfriends in tight, low-cut tops bring in nachos and other goodies. No, I am laying down bedrock that the coming out of life or living things—chemical molecules that could replicate and do things—was good, as is its further evolution. Both life—this way of being—and living things—the lone packages into which life fleetingly puts itself—are *good*.

The first step in ethics is to ask what is good. The *Encyclopaedia Britannica* entry on "Ethics" (1959, pp. 757–758) says, "By 'the good' here is meant what is intrinsically good (or good-in-itself), not what is good only as a means to something else." This is what I mean by the phrase life is good. It is good-in-itself. If there is good-in-itself at all, I would think life is good would be self-evident or unmistakable.

Whether the knowing creation of an Almighty, or the outcome of a wandering, blind, goalless bubbling over of chemistry and electricity in the right setting by happenstance, life and living things are good. Life comes together as neighborhoods in which we as dwellers or wayfarers need to behave as good neighbors to the neighborhood as well as each neighbor. Being a good neighbor is being good to life, which is good-in-itself. The sign at the National Forest[2] trailhead a quarter mile from my front door welcomes hikers, but also warns that we are coming into the home of many kinds of wildlife and that we are "guests in *their* home" (italics on the sign).

By manyfold (manifold) and tangled life, I mean biological diversity or biodiversity. This is the tree of life—many, many kinds of life living in a wealth of jumbled, messy, always-shifting neighborhoods.

By not hobbled by Man's will, I mean *wild*—wild things, which are Earthlings that are as yet self-willed and not thralls to Man.

These other Earthlings are good because they *are* and because they are free by being wild. Wild things are good-in-themselves.

Wild is a manyfold, tangled word and thought. To understand such a word, we need to go back to its beginning in language, at least as far as we can. It means going back to the Anglo-Saxons coming into Britain as Roman civilization was withering and leaving. Early Gothonic or Deutsch speakers—warlords ("kings" and "lords" in their high-and-mighty gall), churls, and bards, such as those who wrote down *Beowulf* along with other sagas and poems—struggled with will. They lived next to wilderness, land not yet settled or plowed, and knew wildlife such as bear, wolf, lynx, wolverine, moose, wisent, eagle owl, snowy owl, golden eagle, white-tailed eagle, and other mighty beings that were untamable. They saw such lands and beings as wild—that is, as having their own will, or as being self-willed—in marked odds with land, crops, livestock, and men under the will of Man or a master.[3] Such lands and waters—wild neighborhoods—called for being "wellthoughten" owing to their freedom and self-worth.[4]

The Anglo-Saxon word for animal is *deor*, which came to mean one kind of animal—deer—after the Normans gave English the word *beast* and Latin gave it the term *animal*. (Deutsch [German] still has *tier* to mean animal, and Swedish has *djur*.) The Anglo-Saxon word for untamable deors was *wildeor* or *wildedeor*—self-willed animal or being. Our word for wilderness comes from here as *wildeorness*—home of wildeors, says Roderick Nash (1967), the world's leading wilderness scholar.

In a 1983 talk at the Third World Wilderness conference in Scotland, philosopher Jay Hansford Vest (1985, p. 324), another thoughtful and careful scholar, also sought the meaning of wilderness in Old English and, further back, Old Gothonic tongues. He believed that wilderness means "'self-willed land' . . . with an emphasis on its own intrinsic volition." He interpreted *der* as "of the," not as coming from *deor*. "Hence, in wil-der-ness, there is a 'will-of-the-land'; and in wildeor, there is 'will of the animal.' A wild animal is a 'self-willed animal'—an undomesticated animal—similarly, wildland is 'self-willed land.'" Vest shows that this willfulness is up against the "controlled and ordered environment which is characteristic of the notion of civilization." These early northern Europeans were not driven to wholly lord over wild things; thus, wilderness "demonstrates a recognition of land in and for itself."

It took the English some 900 years to come up with another Anglo-Saxon word for deor instead of beast or animal. That word is *wildlife*, and it began as *wild life*, then *wild-life*, only about 150 years ago. The need for the word *wildlife* shows, I think, that the terms *wild animal* and *beast* were not fully up to the job. Smithing the word *wildlife* came at the beginning of the conservation uprising in the United States and may have been a beacon for the shift in how some folks were coming to think about other Earthlings.

If wildeors or wildlife are good-in-themselves, then to knowingly or carelessly make another "lifekind" go extinct is evil. Today we are not making only one or two wildeors go extinct, we are bringing on a mass extinction unlike any other happening for 65 million years (Foreman, 2004; Leakey & Lewin, 1995; Wilson, 2002). That last mass extinction was when a comet smacked Earth, and dinosaurs and many other kinds of life were wiped out. Indeed, in the more than 500 million years since the rise of complex animals, there have been only five other extinctions so great as to be called mass extinctions. It is this sixth great extinction that is the upshot of a being so mighty and many as Man *not being a good neighbor*.

For wilderness and wildeors today, then, Man must show restraint—braking its self-willed might—by leaving some lands and wildlife alone, by not stamping its will on them. To be a good neighbor is to hold high the self-will and inborn goodness of one's neighbors, which means sometimes braking one's own will if one is mightier or luckier. This is *good* behavior. Martin Heidegger called on us to "let beings be" (quoted in Devall & Sessions, 1985, p. 99). Leave other beings alone to themselves, to be their own being and follow their own path as cobbled out by evolution, ecology, and happenstance. When we let other *beings* be, we are good neighbors.

Letting being be and letting beings be is *not* what Man has done in its 50,000-year life span so far. Our tale can best be told as *not letting beings be* but rather as enslaving, remaking, and killing other beings. Our chosen job as *Homo sapiens* has been to spread our will through our unstoppable might over other beings and their neighborhoods, to take away their self-will. This has been the core work of hunter and gatherer, tiller and king, priest and philosopher, mother and father, warrior and builder. Not letting beings be is what Man does. Might makes right. This is not very neighborly. By not letting beings be, not seeing other Earthlings as good-in-themselves, we are plundering Earth and all its life, tearing off great limbs from the tree of life and poisoning its roots.

Here, then, is the heart of conservation. Keeping wilderness and wildlife free, hale, and hearty is about letting beings be, about growing the goodness of self-braking that lets land and living beings have their own will. Learning self-braking is key to growing up. Law and ethics call for us to have self-restraint when dealing with others of our kind. Wilderness Areas and many kinds of wildlife can *be* thanks only to our willingness to brake ourselves and our body of selves.

The nineteenth century in North America was an unfettered binge of soil-, sap-, and bloodletting as we scalped land and other beings from the Alleghenies to the Pacific, as we cut down the great tall trees, slaughtered the bison and passenger pigeon, and plowed up the tallgrass prairie—and did our best to kill off the other men, women, and children already living here in widely scattered bands, tribes, and chiefdoms. When conservation stood up in the late 1800s, it was all about braking, holding back, and not doing that which we had the might to do.

When Leopold (1949, p. vii) wrote, "There are those who can live without wild things, and those who cannot," he named who conservationists were. We are *Cannots*—women, men, and children who cannot live without wild things; we are the Cannot Club. This is bedrock. But what are wild things? We've already learned that something wild has its own will or is free of the will of another. So what does the word *thing* mean?

Thing is a Swiss-Army-knife kind of word in Anglo-Saxon and today's English. A wild thing can be a living being—animal, plant, fungus. . . . It can be a neighborhood of wild beings—on land or sea. A wild thing can also be a geologic feature—a river or mountain. It can be weather—a blizzard or flood—and other natural happenings, such as a sunset. Wild things are also ecological and evolutionary processes, such as predation and pollination.

We are now getting to the pith of wildness. Leopold (1949, p. 199) saw wilderness as the theater for "the pageant of evolution." Biologist Michael Soulé, founder of the Society for Conservation Biology and the Wildlands Project, has written much the same thing. Evolution embodies wild things being for their own sakes. Evolution is good-in-itself. In the early 1950s, National Park Service biologists Lowell Sumner and George Collins called for setting aside what is now the Arctic National Wildlife Refuge in northeastern Alaska as wilderness where evolution could go on without meddling from Man. Historian, bush pilot, and Arctic National Wildlife Refuge wilderness manager Roger Kaye (2006, p. 17) writes:

Collins brought to northeast Alaska the belief that the highest experiential values of wild areas derived from understanding that natural processes are ongoing, they are evolving, they are beyond good and bad. They are "right" because they are right unto themselves and can evolve naturally without the medium of man.

Kaye (ibid., p. 21) also notes that

Sumner expressed the hope that this place might always have the "freedom to continue, unhindered and forever if we are willing, the particular story of Planet Earth unfolding here . . . where its native creatures can still have the freedom to pursue their future, so distant, so mysterious."

Evolution is wild. It is wild in the deepest meaning of the word, and thus is the hallmark and highest good of wilderness.

If we are to be good neighbors, if we are to let beings be, and if we are to fit in with other Earthlings for the long haul, then we must step back somewhere (many somewheres) so evolution is free to unfold for wild things in its own unhobbled, eerie way.

The most needed and holy work of conservation is to keep whole the building blocks of evolution along with the sweeping landscapes such as the Arctic National Wildlife Refuge where that unforeseeable, unfathomable wonderwork can play out unhindered. We need to bring evolution back to the fore as the highest good to be shielded by conservation; sadly, it has faded from sight since its heyday in the 1950s.

Such is the true work of conservation, the goal of the Cannot Club.

Part II

So with this bedrock and background, let us turn our heed to how sails the conservation ship today. For it is the conservation ship that must lead the way for finding how we can be good neighbors to all Earthlings.

First we need to understand that from the beginning, there have been and yet are two kinds of folks who call themselves conservationists. As most readers know, and as mentioned above, the nineteenth century in North America was an unbridled slaughter of wildlife and scalping of wilderness. These *landscalpers*, as I call them, had no heed for the morrow, and sought only to skin the land for the most short-term gain and then head on to the riches over the next ridge. They carelessly and greedily stamped Man's will over wild landscapes, and gleefully slew wildeors— forty million bison and over one *billion* passenger pigeons in the score of years after Appomattox. Against this wanton wrecking, a wave called *conservation* swelled in the waning days of the century to stop the landscalpers, bring scientific management to natural resources, and shield Nature, or in other words, wild things.

This conservation rising soon was riven by blowups over logging and livestock grazing in forest reserves (soon to be called National Forests) and dams in National Parks (Nash, 1967). Stephen Fox (1985), in his topnotch book *The American Conservation Movement*, saw the two kinds of conservationists as heirs to either John Muir, founder of the Sierra Club, or Gifford Pinchot, founder of the US Forest Service (USFS). These two camps are widely cleft in how they see wild things, with the Pinchotians driven to stamp Man's will over the whole world so as to manage it for the greatest "production" of resources or raw goods for the greatest number of Men—in a "scientific" and "sustainable" way, mind you.

Conservation, as I see it, is about shielding wildlife, wilderness, and outdoor loveliness (scenery, both "monumental" and "sublime") from scalping and being made ugly, and about keeping lands and waters for solitude and nonmotorized recreation. Its key word is the old standby *Nature*. By Nature I mean wildlife, ecological doings, and wild neighborhoods—evolution in all its wonder and thrill as well as being in the outdoors with wild things. I do not mean the "built environment" mostly made and willed by Man.

Notwithstanding this cleavage, wide in spots, both conservations stood for braking the landscalpers and called for self-restraint in our dealings with Earth.

Paul Shepard (1967, pp. 236–237), a leading conservationist and "human ecology" professor in the second half of the twentieth century, called the resource conservationists *resourcists*. *Resourcism* is a better name for what Pinchot's followers do than is the word *conservation*. Therefore, I use *conservation* for Nature conservation and *resourcism* for resource conservation. I acknowledge that darn few, if any, resourcists would call themselves such; I think most still hold on to being *conservationists* and call my kind *preservationists*, just as Pinchot called Muir one hundred years ago.

Another cleft is how conservation and resourcism see the good of Nature—what I call wild things, but they call natural resources. Here we get into the real meat of the split—the underlying belief gap between conservation and resourcism: conservationists believe in keeping wild things for their own sake. Resourcists believe in managing natural resources for products and uses for humans. *Multiple use* is their watchword. Conservationists believe that many lands and waters should abide wild (free of Man's will); resourcists believe that Man should tame nearly all lands and waters. There is a lot of room between these two beliefs, and much wandering about in between. This is a map, after all. Lincs in the true world are seldom razor sharp.

It is also true that most of the *public* arguments made by wilderness and wildlife lovers are grounded in recreational, aesthetic, utilitarian, and lately, ecological and scientific values. Nonetheless, since at least the time of Henry David Thoreau, many wilderness and wildlife conservationists have spoken out about how they care for wild things *for their own sakes*, seeing them as good-in-themselves. This belief has been stated in many ways. The late Canadian naturalist John Livingston (1981, pp. 17–18) wrote that wildlife conservation is "the preservation of wildlife forms and groups of forms in perpetuity, for their own sakes, irrespective of any connotation of present or future human use." He went on to say, "In essence, wildlife conservation is the preservation of nonhuman beings in their natural settings, unaffected by human use or activity, uncontaminated by human antibiosis, emancipated from human serfdom." In this I believe he spoke for most of us who would not want to live without wild things, whether or not some shier Cannots would put it so bluntly to others.

Set against Livingston, a Bureau of Reclamation cheerleader wonderfully set out the innermost belief of hard resourcism when the heyday of dam building was getting under way:

The destiny of man is to possess the whole earth, and the destiny of the earth is to be subject to man. There can be no full conquest of the earth, and no real satisfaction to humanity, if large portions of the earth remain beyond his highest control. (Widtsoe, 1928, p. 138)

Such is the gap between the two bands that call themselves conservationists. A more enlightened statement of how resourcism *can* bestow worth on wild things comes in a 2007 article in *Science*: "The case for biodiversity conservation can be argued on economic, sociocultural, and aesthetic grounds" (Butler, Vickery, & Norris, 2007). Yet even here there is no acknowledgment that wild things may be good-in-themselves; biodiversity is worthwhile only for what good it offers Man. We hear the same from those who call on conservationists to uphold biodiversity by highlighting how wild things give Man "ecosystem services" and who tell us not to talk about the inborn worth of other beings, as such talk will put off politicians and the public. Such conservationists are in truth a kinder, softer stripe of resourcists, who see biodiversity and ecosystem services as natural resources worth many dollars, just as timber, minerals, water, and other resources or "multiple uses" are worth dollars.

The beliefs of resourcism and conservation come to life in policy as well as on-the-ground management, protection, and restoration. In the twentieth century in the United States and Canada, I believe, federal, state, and provincial resource agencies snuffed and yoked more wilderness than did resource industries, poachers and their ilk, and other landscalpers. In shoving wildlife to the brink, the agencies have been as bad as the landscalpers. Most of the work of conservationists on the public lands (Crown lands and parks in Canada) has been to thwart agency wilderness "development" and wildeor-killing work.

Far from keeping things wild as they should, federal and state/provincial land- and wildlife-managing agencies work mostly to help timber, mining, livestock, energy, commercial fishing, trapping, outfitting, tourism, and other industries rip out raw goods from wildlands (and seas), and build resorts and other developments in wildlands. In Canada, federal and provincial agencies have lately been even worse than their US mates, helping landscalpers slaughter wildlife on land and sea as badly as anywhere in the world today. We must understand that resource agencies have an agricultural mind-set. In some state and provincial wildlife agencies, things are slowly shifting thanks to "nongame" wildlife and endangered species programs along with the kind of "wildlifers" coming in to work for and even run such programs. Elsewhere, through the hammer of political will, state wildlife agencies (such as in Alaska, Idaho, Wyoming, and the Dakotas foremost) have flipped back to the early 1900s to kill off wolves, mountain lions, prairie dogs, and other "pests" (the same is happening in Canadian provinces).

I am weighing here the beliefs and deeds of institutions. Folks working for land and wildlife agencies since the beginning have been widely spread in how they think about and work with wild things. In other words, agency staffs in the United States and Canada have been made up of landscalpers, resourcists, and Cannots. Some have loathed mountain lions, lands without roads, and uncut big woods. Others have done what they think is a good job caring for wildlands and wildlife while overseeing taking raw goods from wildlands. Yet others have taken bold, sometimes-job-ending stands to keep wilderness and wildlife hale, and fight sloppy, over-the-top logging, grazing, commercial fishing, and slob hunting wanted by the economically and politically strong.

Getting new National Parks and Wilderness Areas named by government as well as shielding and restoring threatened wildeors have been at the heart of the struggle between conservationists and resourcists. In the United States, Wilderness Areas are by law federal lands that have no permanent human dwellings, no permanent roads, no use of motorized or mechanized equipment or vehicles (except for emergencies or vital administrative needs), and no commercial logging or other development.[5] Wilderness Areas are on the whole open to nonmotorized, nonmechanized recreation as well as lawful fishing and hunting, and for scientific study.

In the clash over Wilderness Areas (and other wild havens) between the USFS and other land-managing agencies, on the one hand, and wilderness clubs, on the other, the catfights swirl about choices, such as how many new Wilderness Areas or other wild havens should be set aside; which landscapes should be set aside as havens; how much of such landscapes should be kept wild; where should boundaries of such havens be drawn on the ground; how many "cherrystem roads" and other "intrusions" should be left open into the Wilderness; should any dollarable trees or other raw goods be "locked up" in the Wilderness; and should playgrounds wanted by off-road vehicle, mountain bike, and snowmobile thrill seekers be carved out of the Wilderness.[6] These are technical quandaries, yes, but they get to the heart of clashing beliefs about wild things. More often than not, the agencies work for "the less, the better" and "the weaker, the better."

So in a nutshell: conservationists work to keep Man's will from taming all of Earth; resourcists work to stamp Man's will over as much of Earth as they can, even for the wildest wildeors (grizzly bears, wolves, and big cats) and in designated Wilderness Areas through "management" of some kind. *Whose will?* is the bedrock question behind conservation battles.

Part III

With this understanding of the two conservations taken care of, I now wish to give the Muirian clan of conservationists—the Cannots—a Dutch uncle talk. As a whole, Muir's offspring, such as the Sierra Club, Wilderness Society, Nature Conservancy (TNC), and a throng of other international, national, and local outfits, are wealthier and bigger by far than ever before, but also often likely to be more muddled and cowering—a plight that would have irked the matchless Bob Marshall, who at the beginning of the Wilderness Society in 1935, said flatly, "We want no straddlers" (Fox, 1985, p. 211). Now the conservation movement has fence-straddlers galore, and most frequently running things. Some outfits, the once-great TNC foremost among them, have quit straddling the fence and now stand fully over on Pinchot's side—where the big corporate money happens to be. (Some TNC staff members, however, yet stand stalwartly on Marshall's side of the fence.)

My behest to those who, unlike TNC, have not yet become out-and-out resourcists is to quit straddling the fence, and get back on the side of Muir, Marshall, and Leopold (gladly, a sturdy band of clubs and wild lovers still do). To bolster my bidding, I'll call on five kinds of little birds I've come to know well the last ten years. They each have something to tell us. When we learn one of these lessons, we may pick up a cast-off feather from that bird and put it in our Cannot Club hat (it is unlawful as well as bad juju to *take* a feather from a bird). I give these teachings from my feathered friends since we need to learn from them how to better do our job of leading other men and women to become good neighbors to other Earthlings, or at least to set aside sweeping big wild havens where wild things can live free and evolution can work its uncanny *wundorcraeft*.[7]

But let me first explain how I came to know these birds so well. Ten years ago I badly hurt my back in a fall. Notwithstanding surgery that fused two disks, heavy doses of strong pain medication, yoga, core-strengthening exercises, acupuncture, and walking when I can, I've become pretty much a gimp. I can't run or backpack. At times I can walk for two hours or so on trails; as I finish writing this chapter in June 2011, though, I walked for nearly forty minutes with a back brace and trekking poles—the longest I've done in months. I've had to cut back on driving or flying to give talks, or going to conferences. My life has been thoroughly upended by that fall.

Most of my time is in my living room recliner. So that I can still live with wild things, my wife, Nancy, and I have made our yard as welcome and friendly to birds as we can. Being in an older neighborhood only two long blocks away from a 40,000-acre National Forest Wilderness Area (Sandia Mountains), which climbs 5,000 feet above our 5,800-foot-elevation home, has helped. Outside the living room bay window, we have a birdbath and a spread of feeders. I in my recliner and our black cat, Gila, in the bay window are endlessly uplifted and kept awake by our feathered friends—yard birds, yes, but still wild things. I've tallied fifty-eight species in and over my yard. I cannot overstate how thoroughly I need and love these birds.

Thanks to my living room birding blind, I've gotten to know some birds and who they are well. They have taught me much, and I think that they can teach my fellow Cannots much, too.

We Cannots are the ones who for more than six-score years have scrapped for National Parks, Wildlife Refuges, Wild Rivers, and Wilderness Areas. We have fought to keep roads, motor vehicles, logging, dams, and the other works of Man out of wildernesses and wild rivers not yet shut by law to fleecing. We've worked tirelessly for the Endangered Species Act, and to keep it whole and tough against yearly onslaughts by those in Congress who do the bidding of both yahoo and corporate land-scalpers. We have fought all over the world to shield wildeors of every kind wherever they still live, and bring back home those—like wolves, big cats, rhinos, peregrine falcons, and others—that have been killed or driven out of their home turf.

But right now wilderness and wildlife clubs, big and little, national and local, are being undermined and taken over by two crowds: those on Pinchot's side of the fence—the resourcists; and those fence-straddlers at whom Marshall glowered. I call all these undercutters *enviro-resourcists.*

Enviro-resourcism is my rather clumsy word for those foundation staffers, media, organizational, and other consultants, board members, political operatives, and executive directors and other hirelings of conservation clubs who likely can live without wild things, and who are slowly taking over our clubs and the whole conservation network.[8] They have the might to do this through funding. Once conservation was not driven by dollars. Now it is. Enviro-resourcists undercut tough stands, and weaken or elbow aside our lore, the grassroots path, and an outspoken belief in wild things for their own sake. They bring in their stead such Man-saked mush as "ecosystem services," the economic worth of wild things, political pragmatism, leadership by hired "expert" staff and outside consultants, and remaking grassroots clubs into corporate-framed institutions. Enviro-resourcists are a blend of environmentalists and resourcists, not primarily lovers of wild things. Enviro-resourcism is a broad sweep, and not all of those against whom I warn are blameworthy in every way I tally. As I go through what my five birds teach us, what I mean by enviro-resourcism will become sharper, as will my bidding: that we Cannots must take back our wilderness and wildlife clubs along with the overall network, and get them back on Marshall's side of the fence.

Five birds most of all have taught me. They are not those often held up as beacons of idealized virtues, such as eagles or owls. Nor are they bright flashes of many-hued loveliness such as orioles and hummingbirds. Some may think my five birds are drab, but in their behavior and mood they are anything but drab. As I have gotten to know them better, their true grit fairly blazes.

So let's meet them and hear their tweets of wisdom.

Bushtit: Grassroots

Bushtits are tiny, dull, and gray, but lively, lovable, and cheerful in a way that overflows. They swoop through our neighborhood in a throng of a score or more, swarming into a piñon tree, and cleaning it of bugs and caterpillars, then—zoom—they are off in a straggling, chattering dash to a New Mexico locust. They are not seedeaters but rather pack hunters and flesh eaters (bugs have flesh). Were they raven big, bushtits would be the fright of Earth. But they are the soul of winsomeness to me.

The other birds in our neighborhood squabble among themselves and with others. Yet I haven't seen bushtits quarreling. They get along with one another; even in a tight flock, they don't get in each other's way. By being a flock without a readily acknowledged boss, bushtits bring the foremost lessons for understanding the lore of grassroots conservation, and what we need to do to keep it strong and true.

Amateurs and Professionals

Fox (1985) clove early conservationists into *amateurs* and *professionals*. He sees folk conservation clubs as within the amateur tradition (to wit: Muir and the Sierra Club), and government agencies and professional societies (Pinchot and the USFS) as within the professional tradition. The key to understanding the amateur–professional cleavage is that amateurs do conservation work out of a *calling*; professionals do it as a job, career, and for hire. It is not mostly about whether one has a job or is hired for conservation work, though, but *why* one does conservation work: for love or money. Nor is it about how much one knows, how hard one works, or how skillful one is. We "amateurs" have more often than not outworked and outfoxed the agency professionals. Moreover, some professionals working for the USFS and other government agencies are themselves within the amateur tradition, and some who work for

nongovernmental conservation groups may be overmuch in the professional tradition—indeed, they are the ones against whom I am warning.

Those who are lucky to have a job with a wilderness or wildlife club,
whether it is big and national like the Wilderness Society, or little and local like the New Mexico Wilderness Alliance, must always bear in mind
that the work of keeping wild things bears far more weight than one's career. Those hiring others for conservation work should always underline
that our work is a calling, not a career, even though one may do it their
whole life, as I have. If that scares off job seekers—*good!* One's worth in
conservation is not shown by one's paycheck.

Clubs Not Corporate Institutions

A threat undercutting the work of those who need wild things is the prod-
ding toward corporatization and institutionalization in conservation outfits. Wilderness and wildlife keeping should be a grassroots, folk undertaking of kith instead of nine-to-five work in posh offices where there are
sharper eyes for institutional growth than for the hallowed call to keep
wild things wild. Words such as *corporation*, *organization*, and *group*
bring with them the whiff of institutionalization. I would rather think
of the Sierra Club, New Mexico Wilderness Alliance, and others as clubs
instead of organizations so as to underline the high worth of putting the
amateur tradition at the fore. We can see each conservation organization,
then, as a Cannot Club, and the conservation movement as a kinship
of such clubs. Likewise, the words *network* and *web* are better for the
overall gathering than is the term *movement*. Even outfits like mine, the
Rewilding Institute, that don't have members need to walk the amateur
path, too.

For some foundations, board members, executive directors, and organizational consultants, the growth and "institutional effectiveness" of
a conservation group is the main work—even over the work to shield
wildlife or keep wilderness. This upends conservation. It's good to have
a healthy, well-run club, yes, but that is only so it is better at fighting
for wild things. Never should we let institutional/corporate needs stand
above the welfare of wild things. We also need to keep our clubs from
growing for the sake of growing. A smaller, leaner, quicker club is more
likely to be skillful and clever than is a bloated outfit with a big cash
flow that looks good in a ledger. Moreover, working for the lean outfit
helps one keep in mind that conservation work is a calling. In a big, fat,
corporate-framed nonprofit organization, even Cannots can be lured into
institution building and fatter paychecks. Career can take over if all about

one are staffers from the professional tradition and a board made up of corporate executives.

As folk conservationists in the Cannot Club, let us be a flock of bushtits.

Ladder-backed Woodpecker: Doggedness

I once watched a ladder-backed woodpecker in our yard drill into a tree for thirty minutes until she got what she was after. These little woodpeckers work harder than anything I know. They are dogged. When they hear or otherwise know a beetle or grub is under bark or in the wood of a limb, they keep pecking away at it. But they are not locked into one spot. If one drill hole doesn't reach their breakfast, they hammer away at their target from another. They don't just dumbly keep pounding their heads against wood. They stop and cock their heads to see or hear if they are on the right track. They shift if they need to, but they doggedly keep at it.

Long Fight Not Quick Tussle

Polly Dyer of Seattle, now in her nineties, has been shielding the North Cascades since the early 1950s. Though there is now a North Cascades National Park and a batch of Wilderness Areas in the North Cascades, she is still working on more Wilderness Areas with her friends (Manning, 2007). To me, Polly is the living embodiment of US conservation. She is dogged.

Defending wilderness and wildlife takes farsightedness. It may take ten or more years of step-by-step building a network of friends for a wilderness before we can get it set aside as Wilderness by Congress. Some foundations and other funders are shortsighted, however, and want something every two years—a congressional session. Foundations will not underwrite a ten-year campaign for thorough, tough, statewide Wilderness Area bills like some of our big wins before. A two-year dash to get a Wilderness Area bill of some kind, even if it leaves lands more worthy and threatened sitting on the shelf—or worse, open for plundering—is what today's foundations seem to like. They don't seem to understand the downside.

Build the Grassroots

Knowing the land, wildlife, law, and issues are all key. But—to use maybe the nastiest fight going today—keeping big, healthy populations of wolves on the ground is much more about building a big, big network of wolf lovers to raise their howls and scare politicians than it is to know every little technicality of wolf policy. Owing to the backwoods Tea Party folks (even if they live in tony suburbs), slob hunters, and many ranchers, we have to overwhelm politicians with our showing. We can't win the fight for wolves through lawsuits alone, through know-how alone, or by hoping chickenhearted Democratic senators will care enough about wolves or the Endangered Species Act to stand up to the loudmouthed, gun-waving wolf haters. We win the fight for wolves—and every other wildeor or wilderness—by calling forth and heartening wild lovers so that politicians are more afraid of us than of those who fear and hate wild things.

The best conservationists are like ladder-backed woodpeckers. We never, never give up.

Western Scrub Jay: Big Dreams

The corvid family—jays, crows, and ravens—are the smartest birds, and western scrub jays might be the smartest of all. Their recall is staggering. A scrub jay knows where it has hidden upward of a thousand nuts. It goes beyond sheer recall, however. Research shows they have a so-called theory of mind, which means they understand that other beings also think—that there are minds other than one's own. Such fieldwork has found that when a scrub jay hides a nut, but knows another jay is watching, it will go back later and put the nut in another hiding spot when the other jay isn't watching. I put out peanuts for the four jays in my yard and then watched them do this sleight of beak. They not only have to watch to see if one of the other three jays sees where they hide a peanut, they also have to watch for curve-billed thrashers, who gladly scoff up on jay-hidden goodies. There are thousands of peanuts hidden all over my yard, some in the most outlandish spots. Jays don't only think; they think ahead. Conservationists need to do the same.

Shove out the Bookends

The great steps forward in conservation, such as the 1964 Wilderness Act and 1980 Alaska Lands Act, have been thanks to big dreams, not to working within the bookends of what we are told is doable. We win more by widening the frame of what can be thought and done than by being hemmed in by what we are told is workable.

From 1977 to 1979, the USFS undertook the second Roadless Area Review and Evaluation (RARE II), and found 62 million acres of roadless areas in the National Forests. In every National Forest, grassroots wilderness folks ground truthed the USFS inventory by walking thousands of miles of boundaries and so-called cherrystem roads jabbing into roadless areas; looked over, mapped, and photographed roadless areas missed by the USFS; and otherwise found what was wrong with the USFS inventory. Our game plan was to play it by the USFS rules and give the agency the kind of factual input it said it wanted. By the end, our folks knew far more about the roadless areas on the ground than did forest rangers. For the roughly 62 million acres (out of 191 million acres of National Forest land), we bent over backward to find a middle ground, and asked for only the "best" 35 million acres to be put forward to Congress as new Wilderness Areas or additions to existing Wilderness Areas.

On the other hand, off-road vehicle hooligans and take-and-scrape industries didn't bother with thoughtful feedback to the USFS; they stomped their feet and yelled, "No more Wilderness." Conservationists sent in many more letters than the antiwilderness crowd, and many more of us showed up at hearings all over the United States. We got screwed nonetheless, even though our friend Jimmy Carter was president. The USFS and Carter offered only 15 million acres for Wilderness, and most of that was "rock and ice"—strikingly lovely peaks and glacial lakes without timber or other raw goods that could be monetized. Our nationwide network of conservation clubs howled their unhappiness, yet got little more than a pat on the head.

In 1980, RARE II was the main nudge for me and a few others to leave mainstream outfits, such as the Wilderness Society, and start Earth First! The first thing we did was to call for all National Forest Roadless Areas to be shielded from logging, road building, off-road vehicles, and other blighting. In other words, we went back and asked for the whole 62 million acres—not only the 35 million. None of the mainstream conservation clubs backed us. But only twenty years later—in January 2001—the chief of the USFS along with President Bill Clinton, with overwhelming support from those who wrote letters or spoke at meetings, put out the Roadless Area Rule, which overall shielded most roadless areas from road building.[9] Nearly all conservation clubs worked their butts off for the rule. We few first Earth First! folks built a new playing field in 1980 by asking for all National Forest Roadless Areas to be kept roadless. We had no backers then. In only twenty years, though, other conservationists and even the USFS and president were playing on our new field.

Dream! Don't Bog Down Reacting

Much conservation work is taking on one threat after another as they are thrown at us. We must fight these "brushfires." Yet we also need to think ahead and run after our own dreams. The Wilderness Act and eight-year struggle for it is the template. The Wildlands Project in the early 1990s called for a new framework for National Parks and Wilderness Areas. This framework came to be called rewilding. Rewilding stands on three legs. One is that protected areas—or *wild havens*—are the foremost work of conservation, the best tool in our conservation tool box, and best shield wildlife when roadless or nearly so; in other words: wilderness. Two is that in North America outside the far north, no wild havens are big enough for wide-roaming wildeors; so we need to link havens together by wildways for daily and yearly wandering, dispersal (looking for new homes and mates), and now, shifting homes owing to climate change. Three is that ecosystems—wild neighborhoods—will wither and crumble without "highly interactive species in ecologically effective populations," such as big hunters (wolves, cougars, and so forth), pollinators, and landscape crafters like beavers and prairie dogs (Soulé, Estes, Berger, & Martinez del Rio, 2003, p. 1238). When the Wildlands Project first offered this new path, it was knocked by almost everybody, even other conservation biologists. But since the late 1990s, rewilding with wildways and brought-back highly interactive species has become the widely acknowledged path for conservation throughout the world (Foreman, 2004).

Let us be, then, like western scrub jays. Think. Look ahead. Have big dreams. Remake the playing field.

Curve-Billed Thrasher: Toughness

Curve-billed thrashers are among my most-loved Earthlings. Their mad orange eyes notwithstanding, they have a loftiness and steadfastness about them that cows me. It cows the other birds in our neighborhood, too. Though jays, robins, and doves outweigh them, thrashers are the boss birds. They own our yard. I put both peanuts and sunflower seeds in one small tray feeder outside my living room window. I watch it while drinking my first cups of coffee. There might be four brassy jays swooping in and out for peanuts to hide, but when a curve-billed thrasher settles in to munch sunflower seeds, scrub jays sit back and wait.

Ask for What You Want, Not What You Think You Can Get

Some of our leaders want us to be "reasonable" and not over-reach. Politicians want us to "negotiate" with landscalpers, and afterward bring them a "consensus" Wilderness bill they can quickly run through Congress and to the president's desk, and then reap editorial praise. Today, politicians wheedle us to back off. This is at odds with how it was not too long ago.

In the late 1980s, Californians were ready to take their Wilderness Areas/ National Parks framework for the California Desert to Senator Alan Cranston, a Democrat who was not thought to be a stalwart for wilderness. Some wanted to cut it back first. "Let's just ask for four million acres and see if Cranston will go for it." But others said, "No, damn it, let's ask for the whole nine million acres of new National Parks and Wilderness Areas we want. If it's too much, then we can cut it back." The daredevils won and took Cranston the big bill. Thanks to the outpouring of letters he had already gotten for protecting the desert, he said, "Sure, let's do it!" It took a few years, but the California Desert Protection Act became law with nearly everything we wanted and still stands as one of the greatest landmarks in Wilderness lore.

Don't Sell Out to Be "Politically Realistic"

On more grounds than I can list here, the conservation and environmental networks have become tied to the Democratic Party. One of the worst outcomes ever of this link to the Democrats was the way some conservation leaders turned their backs when the US Congress in spring 2011 took a sledgehammer to the Endangered Species Act of 1973. Rabble-rousers in the northern Rocky Mountains have whipped slob hunters and ranchers into a lather about how reintroduced wolves will wipe out elk herds and drive ranchers out of business. Out-and-out lies and farfetched overstatements, yes, but they play well with many who dislike cultural shifts in the West.

The big political reality today, we are told by the enviro-resourcists, is that we have to keep the landscalpers in the Republican Party from taking over the Senate in the 2012 election—an outcome that would truly be dreadful. A threatened Democrat is Senator Jon Tester of Montana. Tester thought he needed some raw meat to throw the wolf-hating yahoos in his state, so he sponsored legislation taking the wolf in Montana and Idaho off the Endangered Species Act list, and handing its management over to the states. Such state-run oversight means slaughtering wolves by "sport hunting" down to where they are but a whisper in the wilderness. Tester is a farmer and rancher himself, and likely doesn't care much for wolves. Still, he led the greatest inroad against the Endangered Species Act ever. Politicians who will do damn near anything for re-election are nothing new. But what links Tester's undercutting of the act to our tale is that a few weeks after his strike at one of the keystones of US conservation, the League of Conservation Voters held a high-ticket fund-raiser for Tester in Washington, DC. Among the thousand-buck-a-head "hosts" for the fund-raiser was the head of the Wilderness Society.

True conservationists are seething about how some conservation leaders winked at this shameful deed; it well could end up hurting some Democratic politicians and Potomac enviro-resourcists more than if Tester had
not taken an ax to the Endangered Species Act. Already the League of
Conservation Voters and Wilderness Society have lost a handful of members and donations.

Don't Be Cowed by Taboos

Thirty years ago, nearly all conservation, environmental, and even resource outfits saw Man's population explosion as the underlying driver of
wounds to wild things. Then an awful backslide happened. Talking about
overpopulation, for grounds I explore in my new book, *Man Swarm and
the Killing of Wildlife*, became politically incorrect (Foreman, 2011).
Most foundations backing conservation clubs would not fund population work or even other work if a club talked tough on overpopulation.
Bowing to such a taboo, though, does not make the population explosion
go away. It is only getting worse. We must be bold, speak the truth, and
work on the true nightmares—even if they frighten others. Even if they
frighten away some of our funders.

Quoth the thrasher: "Don't be a chickenshit."

Mountain Chickadee: Own Sake

I've never heard anything happier and merrier than "chick-a-dee-dee-dee."
When these snazzy little gray birds with the sharp black stripe through
their eyes show up and tell all the world "chick-a-dee-dee-dee"—"We are
here!"—I can't help but smile and "chick-a-dee-dee-dee" back at them.
Sometimes I spend the next fifteen minutes softly singing "chick-a-dee-
dee-dee." No wonder my wife thinks I'm a bit daft (and tone deaf).

Mountain chickadees have a good time. Why? Because they live for
themselves. They don't see themselves as a mirthful show for me; they
don't see themselves as any kind of good or help for Man. No, they are
chickadees, and that is all they need to be to answer for their lives and
what they do.

Keep Goodness Uppermost, Not Dollars

There are some well-meaning fence-straddlers, often with PhDs, who believe that the only way to save wild things is to show that they are worth
money. One ballyhooed way is that chemicals we might find in plants,
insects, or salamanders might help us come up with new pharmaceuticals—so-called bioprospecting. Native peoples and third world countries

are thrashing out deals with big pharma about how to split gains from so-called traditional knowledge along with the wild things living in their turf.

Some conservationists have rightly shown how much wolves are worth to the local economy near Yellowstone National Park. Lots of old-time wolf haters howled that if wolves were loosed in Yellowstone, it would scare away all the park goers, and wreck the economies of Wyoming, Montana, and Idaho. They were dead wrong, although they will never acknowledge it. More folks come to Yellowstone now to see wolves than for anything else.

Ecotourism (wildlife viewing, reef diving, birding, and so on) has helped some third world countries economically. Belize, in Central America, is one nation for which ecotourism is a mainstay of the national economy. All of these economic goods can be worthwhile for helping to keep wild things if they are true, not overstated, and if they don't stand as the only grounds for shielding the wild. Belize is good here. The Belize Audubon Society has posters with toucans or other wildlife saying, "Belize is my home, too." Such cleverness gets own-sake goodness understood in a soft kind of way.

But lately, some leading enviro-resourcists and conservation biologists have called for biodiversity arguments to be grounded in what they call "ecosystem services"—such as how forests clean the air and make oxygen. They have reckoned sky-high economic benefits for these ecosystem services. They also say that the ecosystem-services basis should elbow aside any talk about wild things for their own sake and that selfish grounds are the only ones that work for most Men.

Long ago, Leopold (1949, p. 210) warned that if we make conservation "easy," we make it "trivial." He also wrote:

One basic weakness in a conservation system based wholly on economic motives is that most members of the land community have no economic values. . . . When one of these non-economic categories is threatened and if we happen to love it, we invent subterfuges to give it economic importance.

Leopold (1953, p. 150) went beyond that thought, though, when he noted that economics wasn't the why for conservation anyway, it was "whether a hawkless, owl-less countryside is a livable countryside."

In his matchless *The Arrogance of Humanism*, David Ehrenfeld (1978, p. 210) gives over one whole chapter, "The Conservation Dilemma," to the dollar worth of wild things:

I have tried to show in this chapter the devilish intricacy and cunning of the humanists' trap. "Do you love Nature?" they ask. "Do you want to save it? Then tell us what it is good for." The only way out of this kind of trap, if there is a way, is to smash it, to reject it utterly.

It is OK to talk about how keeping wild things is good for the economy when such is true, but only when we underline that economics are not *why* wild things should be kept. The why is that they are good-in-themselves.

And so the mountain chickadees popping into and out of our neighborhood carry the most worthwhile teaching of all for us: they are there for themselves. As are all other wild Earthlings. They are for their own sake. We do not need to weave knotty, tangled philosophical theories on how wild things might be good-in-themselves. Chickadees tell us so. Chickadees laugh in our mugs at the outlandish gall that only we—the hairless ape—can give something worth.

"Chick-a-dee-dee-dee!"

It means that wild things are good for their own sakes.

And when we sing back "chick-a-dee-dee-dee!" it means we have the wisdom, generosity of spirit, and greatness of heart to let beings be.

"Chick-a-dee-dee-dee!"

Conclusion

Let me wrap up this chapter by setting out again my key thoughts. Herein I have dealt out two questions and have done my best to answer them. One was, "How do we fit in and live with all the other Earthlings for the long haul?" Our first step is to acknowledge all other Earthlings as kindred, and understand that we as women, men, and children and as Mankind must learn to be good neighbors to wild things, which are Earthlings that are as yet self-willed and not thralls to Man. These other Earthlings are good because they *are*, and because they are free by being wild. Wild things are good-in-themselves.

Today, Man takes an overweening slice of Earth—we gobble up far more than our share of every neighborhood in our world. If we acknowledge wilderness and wildeors as good-in-themselves, though, Man must show restraint—braking our self-willed might—by leaving some lands and wildlife alone, by not stamping our will on them. Braking ourselves lets evolution flow free. As I earlier wrote, *Evolution is wild*.

The second question is how do we who love wild things—we Cannots—do a better job of showing Mankind that we must brake our will so that wild things, wild neighborhoods, and evolution remain free? And how can we be better fighters (not too strong a word) for wild things and evolution? I have gone to five little birds because I believe if we listen to and watch them, we will learn what we need to do—how to brake the will of all Mankind to leave enough of everything for everyone else.

So I do talk to the birds and all the other wild things I meet, whether at home or on the trail. They are my neighbors. Talking to them is the neighborly thing to do.

Talk to your neighbors. In their tongues and ours.

"Chick-a-dee-dee! Good morning, pretty thing."

It's good for you.

It's good for them.

And it's good for the wild world.

Notes

In the forthcoming books in the For the Wild Things series by the Rewilding Institute and Raven's Eye Press, I will delve into these cares much more deeply than I have room to do here. See http://rewilding.org/rewildit.

1. I use the word *Man* or *Men* capitalized for the species *Homo sapiens*, *woman* for the female of the species, and *man* lowercased for the male. This is more in keeping with earlier English, which had another word for male *Homo sapiens*: *wer*, which lives on today as werewolf. Today's English is odd for a modern tongue not to have a straightforward word for our kind that is also not the gendered word for the male. To have to call ourselves by a Latin word, human, is cumbersome and abstract. I do not write Man in a sexist way but rather for the goodness of the English tongue.

2. I capitalize Wilderness Area, National Forest, National Park, and such to show they are special, legal, land classifications. For example, a wild, roadless landscape that has not been designated as a Wilderness Area by congressional legislation signed by the president is wilderness in lower case.

3. A man without a master was free, or a *freeman*. Likewise, we may think, *deors* (animals) and worts (plants) that were not domesticated or tamed were also free—or wild.

4. Honored or respected in Old English.

5. Some states—foremost New York, California, and Michigan—have designated Wilderness Areas on some state-owned lands.

6. Although I write Wilderness Areas, I mean all kinds of set-aside wild havens such as National Parks and Wildlife Refuges.

7. "Wonderwork" or "miraculous craft" in Old English.

8. I don't have room to do "nuance" here; there are good folks in each of these cubbies who are true Cannots, rooted on Marshall's side of the fence.

9. For eight years, the Bush administration did its best to weaken or get rid of the Roadless Rule, and open all National Forest lands to feckless motorized play and ripping out raw goods. We fought them every step of the way, and now ten years later the Roadless Rule holds overall, although it is still tangled.

References

Butler, S. J., Vickery, J.A.G., & Norris, K. (2007). Farmland biodiversity and the footprint of agriculture. *Science, 315*, 381–384.

Devall, B., & Sessions, G. (1985). *Deep ecology*. Layton, UT: Peregrine Smith.

Ehrenfeld, D. (1978). *The arrogance of humanism*. New York, NY: Oxford University Press.

Ethics. (1959). In *Encyclopaedia Britannica* (Vol. 8). Chicago, IL: William Benton.

Foreman, D. (2004). *Rewilding North America: A vision for conservation in the 21st century*. Washington, DC: Island Press.

Foreman, D. (2011). *Man swarm and the killing of wildlife*. Durango, CO: Raven's Eye Press.

Fox, S. (1985). *The American conservation movement: John Muir and his legacy*. Madison, WI: University of Wisconsin Press.

Gould, S. J. (1991). Preface: Reconstructing (and deconstructing) the past. In S. J. Gould (Ed.), *The Book of Life* (pp. 6–21). New York, NY: W. W. Norton.

Kaye, R. (2006). *Last great wilderness: The campaign to establish the Arctic National Wildlife Refuge*. Fairbanks, AK: University of Alaska Press.

Keeley, L. H. (1996). *War before civilization: The myth of the peaceful savage*. New York, NY: Oxford University Press.

Leakey, R., & Lewin, R. (1995). *The sixth extinction: Patterns of life and the future of humankind*. New York, NY: Doubleday.

LeBlanc, S. A., & Register, K. E. (2001). *Constant battles: The myth of the peaceful, noble savage*. New York, NY: St. Martin's Press.

Leopold, A. (1949). *A sand county almanac: And sketches here and there*. New York, NY: Oxford University Press.

Leopold, A. (1953). Conservation. In L. B. Leopold (Ed.), *Round river: From the journals of Aldo Leopold* (145–157). New York, NY: Oxford University Press.

Livingston, J. A. (1981). *The fallacy of wildlife conservation*. Toronto, Canada: McClelland and Stewart.

Manning, H. (2007). *Wilderness alps: Conservation and conflict in Washington's North Cascades*. Bellingham, WA: Northwest Wild Books.

Mayr, E. (2001). *What evolution is*. New York, NY: Basic Books.

Nash, R. (1967). *Wilderness and the American mind*. New Haven, CT: Yale University Press.

Shepard, P. (1967). *Man in the landscape: A historic view of the esthetics of nature*. College Station, TX: Texas A&M University Press.

Soulé, M. E., Estes, J. A., Berger, J., & Martinez del Rio, C. (2003). Ecological effectiveness: Conservation goals for interactive species. *Conservation Biology, 17*(5), 1238–1250.

Stanford, C. B. (1999). *The hunting apes: Meat eating and the origins of human behavior*. Princeton, NJ: Princeton University Press.

Vest, J.H.C. (1985). Will-of-the-land: Wilderness among primal Indo-Europeans. *Environmental Review, 9*(4), 323–329.

Widtsoe, J. (1928). *Success on irrigation projects*. New York, NY: John Wiley & Sons.

Wilson, E. O. (2002). *The future of life*. New York, NY: Alfred A. Knopf.

10

The Rewilding of the Human Species

Peter H. Kahn, Jr., and Patricia H. Hasbach

As a species, we came of age on the savannas of East Africa and lived a life more wild than we do today. Much of that wildness exists still within the architecture of our bodies and minds, and needs to be rediscovered, re-engaged, developed, and lived—we need to be rewilded—for us as a species to flourish.

That is the short of our position. To make good on it, or least begin to, in this chapter we develop five overarching ideas. First, we seek to reinstate the importance of the primal self in relation to the natural world—especially those parts that are big, untamed, unmanaged, not encompassed, and self-organizing—and the importance of the primal self in relation to other humans. To do the latter, we suggest that as people today, we need to value and find healthy expressions of our primal passions, sexual and aggressive. Second, we argue that we fear our fear of nature too much, and underappreciate how fear of nature can be life enhancing even when it is in response to objective dangers. Third, we need to move beyond the "tepid pool," and welcome greater variation and periodicity in the satisfaction of human needs and desires. Fourth, in terms of constituting social systems, our rewilding will occur more fully within flatter social hierarchies. As a case in point, we will suggest that hierarchical inequities that have emerged within patriarchal cultures— that is, within most cultures today—arose comparatively recently in our evolutionary history, during the time of early agriculture, and perverted feminine (and masculine) first principles around power, knowledge, and sexuality. Finally, we draw on recent research on the psychological effects of technological nature—technologies that mediate, simulate, or augment the human experience of nature—to argue two points. One is that while technologies often benefit us as a species, they have costs that are not so readily apparent, especially when it comes to our encounter with the wild. The second is that being immersed in future technological substitutions of wildness will not likely lead to similar outcomes as occurs when encountering the actual wild.

Some essays, especially those that are of a formal nature, require first and foremost a structured argument, and rise or fall based on their sheer logic and argumentation. This chapter is partly that, and we have sought a structured contention that provides justification for the rewilding project and a systematic, albeit initial, account of the rewilding process. But the content of rewilding is itself frequently primal and emotional in nature. Sheer logic cannot make the argument, especially because the argument depends on our recognizing aspects of our essential natures that have been diminished and distorted. For that reason, we have drawn on three luminous literary figures—William Faulkner, James Baldwin, and John Milton—to help provide wild images, wild ideas, and primal energy.

We seek to offer a proactive agenda. We hope to show that it—the rewilding of the human species—is as practical as it is future oriented, as important to our well-being as many of the advances of our scientific and technological world.

The Primal

Merriam-Webster's online dictionary refers to the term *primal* (2011) as "first in time," "original," "first in importance," "chief," and "primary." Primal accounts, in this use of the term, are central across many disciplines. For example, the big bang represents a primal cosmological account of our universe. It occurred first in time and offers the primary explanation of how the universe formed. Or natural selection represents a primal account of how life evolved on our planet. The mechanism is said to have existed first in time, and been the chief and primary reason for the evolution of species. What, then, is primal in our human relationship with nature? Answering this question begins to help us understand what is wild in our nature, and potentially still primary and of chief importance within us still.

As with the cosmological and evolutionary sciences, it is difficult to gain access to what happened at an earlier point in time. This is especially true when it comes to understanding the human mind and human emotions. Carbon dating, for example, can tell us something of when specific tools were used at specific geologic sites. But it is more difficult to know what the human experienced who used those tools. Another method is to examine accounts of indigenous hunter-gatherer tribes that had existed until recently with little or no impact from modern societies. The idea here is that their way of life represents our life at a much earlier point in time. Later in this chapter, we will draw on both methods to develop a wider account of the rewilding of the human species. To begin, however, we would like to convey, through Faulkner's (1961) writing in *The Bear*, some possibilities of what the primal relationship with nature looks like.

The story involves a boy's coming of age as a hunter, as a man, in "the wilderness, the big woods, bigger and older than any recorded document" (ibid., p. 185). Each season, the men would set up a hunter's camp in these woods. The boy learned how to walk the woods and not get lost, and to track animals. With the men he "ate strange meat, venison and bear and turkey and coon . . . cooked by men who were hunters first and cooks afterward; he slept in harsh sheetless blankets as hunters slept." The son of a Negro slave and a Chickasaw chief, Sam, wise and inscrutable, taught the boy. When the boy was ten, the first time with a deer in range, Sam said, "Now . . . slant your gun up a little and draw back the hammers and then stand still." The boy did this, "but it was not for him, not yet. The humility was there; he had learned that. And he could learn the patience." After some moments, Sam said, "Now let your hammers down" (ibid., p. 190). Sam was pleased that the boy had known, too, that the moment to shoot had passed. The boy then mused, "Anyway, it wasn't him. . . . It wasn't even a bear" (ibid., p. 191).

The story then becomes about not just any bear but rather *the* bear. Old Ben.

Tremendous, red-eyed, not malevolent but just big, too big for the dogs which tried to bay it, for the horses which tried to ride it down, for the men and the bullets they fired into it; too big for the very country which was its constricting scope. (ibid., p. 187)

The bear was "absolved of mortality" (ibid., p. 197). The bear becomes the embodiment of the untamed, unmanaged, and not encompassed, part of the more than human world that can never be known, not fully. The bear found the hunters within his time and purpose, and moved through their space as he so deemed.

During one season, on his way to manhood, the boy sought to find the bear. "'I will have to see him,' he thought. . . . 'I will have to look at him'" (ibid., p. 198). He ranged the summer woods with his own gun, a new breechloader. He found the bear's print, and returned to camp later and later each day, unsuccessful. "'You aint looked right yet,' Sam said. . . . 'Likely he's been watching you.'" (ibid., p. 199). The boy was taken aback. Then "in a peaceful rushing burst," the boy said, "All right. Yes.

But how?" Sam responds, "It's the gun. . . . You will have to choose" (ibid., pp. 199–200). The boy left the next morning before light, without breakfast.

He had left the gun; by his own will and relinquishment he had accepted not a gambit, not a choice, but a condition in which not only the bear's heretofore inviolable anonymity but all the ancient rules and balances of hunter and hunted had been abrogated. (ibid., p. 200)

By noon he was far into "new and alien country." He navigated with a compass and "the old, heavy, biscuit-thick silver watch which had been his father's." After nine hours, he stopped for the first time. He knew he could no longer get back to camp by nightfall. There was still no sight of the bear.

He stood for a moment—a child, alien and lost in the green and soaring gloom of the markless wilderness. Then he relinquished completely to it. It was the watch and the compass. He was still tainted. He removed the linked chain of the one and the looped thong of the other from his overalls and hung them on a bush and leaned the stick beside them and entered it. (ibid., p. 201)

After several more hours, the boy realized he was lost. Then beyond a log he saw a little swamp, and

he did what Sam had coached and drilled him as the next and the last, seeing as he sat down on the log the crooked print, the warped indentation in the wet ground which while he looked at it continued to fill with water until it was level full . . . and the sides of the print began to dissolve away. Even as he looked up he saw the next one, and, moving, the one beyond it; moving, not hurrying, running, but merely keeping pace with them as they appeared before him as though they were being shaped out of thin air just one constant pace short of where he would lose them forever and be lost forever himself, tireless, eager, without doubt or dread . . . emerging suddenly into a little glade; and the wilderness coalesced. It rushed, soundless, and solidified—the tree, the bush, the compass and the watch glinting where a ray of sunlight touched them. Then he saw the bear. It did not emerge, appear: it was just there, immobile, fixed . . . not as big as he had expected, bigger, dimensionless against the dappled obscurity, looking at him. Then it moved. It crossed the glade without haste, walking for an instant into the sun's full glare and out of it, and stopped again and looked back at him across one shoulder. (ibid., p. 202)

The boy had to relinquish control. That is primal. Technologically, he had to leave his gun behind, and then his compass and watch. With them, he would always have known where he was. But he would never have gotten to where he was seeking. He needed to be lost in order to be found. That's primal, too. The boy needed to relinquish the need to dominate the natural world: to control, trample, and kill it.

The technologies of any civilization can be its trappings. Perhaps most of us know that to some degree. Many people in urban settings, for example, leave behind the comforts and safety of their confines, if only for a day, to hike into the woods and experience a small part of the boy's experience. Some people camp outside, and relinquish the safety and comfort of a tent, if only for a single night. That decision to do so allows sensorial access to the night sky. It is a primal encounter. The night sky that we see has existed for as long as there has been a sun and an earth. Sometimes people, like the boy, go alone into nature, recognizing that the solitude itself provides another means of relinquishing control and opening oneself up to the primal Other. John Muir (1976, p. 314) put it this way: "Only by going alone in silence, without baggage, can one truly get into the heart of the wilderness. All other travel is mere dust and hotels and baggage and chatter."

The primal exists in our relation to external nature. But humans are part of nature, too. We are human nature, and thus the primal also exists within us and in our relation to one another. Two powerful primal forms of energy involve sexuality and aggression, and even when distorted, they can sometimes provide a starting point for tapping into something deep and powerful and ultimately beautiful within the human psyche.

As a case in point, here's a scene from a novel, *Another Country*, by Baldwin (1960). A black man is at a party during the nighttime in New York City. He has started a relationship with a white woman. They are on a balcony.

Her fingers opened his shirt to the navel, her tongue burned his neck and his chest; and his hands pushed up her skirt and caressed the inside of her thighs. Then, after a long, high time, while he shook beneath every accelerating tremor of her body, he forced her beneath him and he entered her. For a moment he thought she was going to scream, she was so tight and caught her breath so sharply, and stiffened so. But then she moaned, she moved beneath him. Then, from the center of his rising storm, very slowly and deliberately, he began the slow ride home. . . . And, shortly, nothing could have stopped him, not the white God himself nor a lynch mob arriving on wings. Under his breath he cursed the milk-white bitch and groaned and rode his weapon between her thighs. She began to cry. *I told you,* he moaned, *I'd give you something to cry about,* and, at once, he felt himself strangling, about to explode or die. A moan and a curse tore through him while he beat her with all the strength he had and felt the venom shoot out of him, enough for a hundred black-white babies. (ibid., pp. 21–22)

What is it exactly that makes it primal? Maybe some of it is the violence? That is part of the Marlon Brando character in *A Streetcar Named Desire.* Maybe that sexual violence—and a man using a woman, and a woman using a man—is part of our evolutionary history. We do not know. We do not know if anyone knows. Regardless, today that violence may well be a perversion of the primal. But that energy is not.

The sexual primal energy without violence and domination—what does it look like? Maybe the answer has embedded within it still the concept of *use*, but of a different form.

According to Richard Nelson, Koyukon elders of their traditional way believe no animal should be considered inferior or insignificant. Each animal deserves respect. All are part of a living community. It is a community that includes not only humans and animals, and not only plants, but also mountains, rivers, lakes, and storms—the earth itself. Nelson (1989, p. 13) writes:

> According to Koyukon teachers, the tree I lean against feels me, hears what I say about it, and engages me in a moral reciprocity based on responsible use. In their tradition, the forest is both a provider and a community of spiritually empowered beings. There is no emptiness in the forest, no unwatched solitude, no wilderness where a person moves outside moral judgment and law.

In the way of the Koyukon, in a primal way, one kills. One kills to eat. But one kills with respect, engaged in a moral reciprocity based on responsible use.

Imagine, then, a person *using* another person, yet based on a moral reciprocity that is itself based on responsible use. What might that look like? Perhaps it is first feeling confident in giving the other pleasure, and when that is substantiated and solidified psychologically the "use"—the "objectification"—of the loved one can occur, and in that context the other welcomes it, needs it, and that need drives the other forward as much as the forward motion drives the other, and the two become one within the one and within the other, and—as Baldwin says—there is no stopping it. Perhaps that is Baldwin's primal energy done well.

Fearing Nature

As part of rewilding, we have so far suggested three overarching ideas. First, as a species we need to value that which is primal in the natural world, including that which is big, vast, untamed, unmanaged, not encompassed, and mysterious. Second, we need to rediscover primal ways of accessing this world, such as relinquishing control and leaving aside our technological trappings. Third, we need to value and find healthy expressions of our primal passions. We now move another step forward in our account of rewilding. In this section, we want to suggest that to rewild, we need to open ourselves up to fearing nature in an appropriate way.

When people speak of the importance of nature in their lives, they often emphasize beautiful experiences. Perhaps they speak of gardening, picking huckleberries on a lazy summer afternoon, walking along the seashore, romping with one's dog, or having a family picnic in a nearby park. Happy experiences all. But in terms of a fuller account of the human relationship with nature, there is a harsher side. Holmes Rolston (1989, pp. 49–50) puts it this way:

[Nature] remains our life partner, a realm of otherness for where we have the deepest need. [Yet] I resist nature, and readily for my purposes amend and repair it. I fight disease and death, cold and hunger. . . . Environmental life, including human life, is nursed in struggle; and to me it is increasingly inconceivable that it could, or should, be otherwise. If nature is good, it must be both an assisting and a resisting reality. We cannot succeed unless it can defeat us.

As Rolston says, we amend nature. We resist it. We struggle and fight against it.

We can and at times should also fear nature. Such fear can save our lives. It keeps us from encountering unnecessarily dangerous situations, and helps protect us when we do encounter them. Catch a glimpse of a rattlesnake from the corner of your eye? Quickly our heart rate increases, adrenaline enters the system, and we are in a position to respond effectively, in a manner often referred to as "flight or fight."

Research supports the proposition that people respond differently to fearful aspects of nature compared to fearful aspects of the human-created world. In one laboratory experiment, for example, people were presented with images of three types of entities at a rate faster than the conscious mind can identify. Results showed that people responded physiologically (e.g., with increased heart rate) when presented with images of dangerous animals, but not when shown images of dangerous artifacts or neutral stimuli (for reviews, see Gullone, 2000; Ulrich, 1993).

This distinction between types of fear was explored further in a study on children's conceptions of bats (Kahn, Saunders, Severson, Myers, & Gill, 2008). In that study, 120 children were interviewed after each child had exited a zoo exhibit that housed Rodrigues fruit bats. The study was centrally focused on the question of whether children could care for an animal they also feared. The answer was yes. The results showed, for example, that the majority of children said they cared about bats, liked bats, would care if there were no bats in the world, and it would matter to them personally if they lived their whole life without ever seeing a live bat, in the zoo or the wild. That said, almost three-quarters of the children said they would prefer not to sleep in a place where bats were able to fly around, and about half said that in such a context, they would have some fear. Thus, the study established that in the same children, a fear orientation toward bats existed alongside a caring orientation. Within the context of this study, two specific interview questions are particularly relevant for the purposes of this discussion. Children were asked:

One person I talked with said that the fear she (he for male participant) feels with bats is similar to the fear she feels when walking down a dark city street at night. Do you feel this way? Why or why not? (ibid., p. 379)

The results showed that of the children who had said previously they felt scared or nervous with bats flying around, 44 percent said that the two types of fear (the fear of bats versus the fear of walking down a dark city street at night) were different. The children were then provided with an alternative account of fear of bats—one where that fear of nature was welcomed. The question was asked as follows:

Another person I talked with said that the fear she feels with bats is very different from the fear in the city. Rather, she said she kind of likes the fear she feels with bats. Do you feel this way? (ibid.)

The results showed that of the children who had said previously they felt scared or nervous with bats flying around, 45 percent said they "kind of liked" the fear they felt with bats.

Hence, there is some initial convergence of empirical evidence for the proposition that people's fear of nature differs from people's fear of dangerous artifacts and urban life; and people can experience their fear of nature as wholesome, life affirming, and sometimes even enjoyable.

We have found it so in our own lives. We spend part of the year in a remote mountainous area, and occasionally encounter rattlesnakes, bears, and scorpions. The encounters are meaningful, and their memories can last a lifetime. But they do not happen often. Rather when walking the land, pulling wood from the woodpile, or going to the outhouse, we stay mindful of our surroundings. We find our mindfulness restorative while in motion. And on our return to the safety of the cabin, we feel more relaxed, poised, and mentally rested and alert.

It is true, of course, that nature offers real dangers. Some of them can kill people. Some do. Rewilding involves facing such dangers through strength, confidence, and even at times some bravado.

In her book *The Old Way: A Story of the First People*, Elizabeth Marshall Thomas (2006) recounts her several years of living with a hunter-gatherer group in the 1950s. This group, the Ju/wasi of the Kala-hari Desert, had at that point experienced virtually no contact with West-ern people. Thomas writes that when the Ju/wasi had chance encounters with wild predators, such as hyenas and lions, they did not present their backs to the animals and certainly did not run. Instead, they sought to ap-pear "indifferent and confident" (ibid., p. 143). Thomas recounts the sto-ry of a Ju/wa woman caught all day in a metal-jawed animal trap set by a white man. For the entire day, alone on the savanna, bleeding, trapped, and in pain, the woman stood tall and strong; to show fear would be to invite death. Thomas (ibid., pp. 150–151) also shares the following tale:

I remember one night in particular. . . . I soon heard a man's strong voice in a stern, commanding tone telling someone to leave immediately. The Ju/wasi never took that tone with one another. I came out of the tent to see what was happen-ing, and behind some of the shelters I saw four very large lions, each three times the size of a person. . . . The speaker was ≠Toma. Without taking his eyes off the lions, he repeated his command while reaching one hand back to grasp a flam-ing branch that someone behind him was handing to him. He slowly raised it shoulder-high and shook it. Sparks showered down around him. "Old lions," he was saying firmly and clearly, "you can't be here. If you come nearer we will hurt you. So go now! Go!" . . . The lions watched ≠Toma for a moment longer, then gracefully they turned and vanished into the night.

≠Toma presumably experienced fear. Yet that fear did not prevent him from acting firmly, strongly, and confidently.

While there is a healthy primal fear of nature, we hasten to add that in our urban and technological world, we too frequently think nature is much more dangerous than it really is. For example, at a large public beach in Northern California, there is a sign that conveys the follow-ing message: "Warning, a shark attacked a person in 1998 fifty yards offshore. These waters can be dangerous." The sign invokes fear, and in our view unnecessary fear. It prevents some people from swimming, and orients other people who enter the water to engage in an almost para-noiac encounter. As a point of contrast, how many physical attacks by humans on other humans have taken place in the beach's parking lot since 1998? No sign says. Probably many. But we do not read signs that say: "Warning, a person was attacked in this parking lot in 1998. This parking lot can be dangerous." It is something about the wild aspects of nature that we hold out as particularly threatening. A similar situation occurs with mountain lions. Attacks are seldom. The paranoia is high. With the paranoia comes the drive to eradicate the source of the fear: we kill the animal) Other. We eradicate wild species. In so doing, we damn ourselves to encounters with the "Others" who are merely reflections of ourselves.

The Tepid Pool

In a New Jersey town, a controversy about a pool pitted neighbor against neighbor. Here was the issue, as reported in the *New York Times* (Applebome, 2009). There is a natural swimming hole called Graydon Pool. It is 2.6 acres. For over ninety years, children in that town have grown up swimming in this pool. It has a sandy bottom. Cool spring currents flow into it. Many residents, though, wanted to plow under this natural pool and replace it with a blue, concrete pool with "thoroughly disinfected" chlorinated water. They called this a "real pool"—a "bona fide pool." But if that happens, imagine what would be missed by the future generations of kids in this community who would not get the chance to swim in the natural swimming hole—the feeling of the sand between their toes, the fluctuations of water temperature in places where the cool streams feed into the pool, the periodicity of usage due to the seasonal fluctuations, the presence of bugs that might land on the water and birds that might be on the shore, and a connection to a natural ecosystem. Children will lose these experiences and will not even realize what it is they have lost.

Toward rewilding, we would like to make a case for greater variation and periodicity in the satisfaction of human needs as well as desires than what is considered normal today. Such greater variation and periodicity was so in our ancestral times. For instance, after a successful hunt the Bushmen could have large quantities of meat for several days. At other times there was little meat. Lorna Marshall (1976, p. 177; quoted in Thomas, 2006, p. 102) writes: "The intense craving for meat, the uncertainty and anxiety that attend the hunt, the deep excitement of the kill, and finally, the eating and the satisfaction engage powerful emotions in the [Bushmen] people." Water was usually scarce for the Bushmen. When hunters set out, "they would drink prodigious amounts of water, until their stomachs were bulging, and would do this again on returning, and in between would choose water over food if the two sources were not together" (Thomas, 2006, p. 95). The Bushmen lived in a desert with few water holes. Still, when the rains came, there would be brief times when the desert pans filled with water, and the children and adults could swim. It was a time of great happiness. So it was for many hunter-gatherer populations (Shepard, 1993, 1998). There were times of activity, hunting, and foraging, of course, but they were balanced by times of leisure around camp, such as storytelling around a fire ring while food cooks slowly, beading, and playing music on their hunting bows. Hunter-gatherers were responsive to sunrise and sunset, the migration of animals, when berries would be sweet, roots ready for digging and nuts ready for harvesting, clouds forming, seasons changing, heat and cold, and predators resting.

In ancestral times, we learned how to thrive and experience deep pleasures within these rhythms and constraints imposed by a demanding yet opulent natural environment. Now through our agricultural and technological capabilities, however, we diminish as well as buffer ourselves from nature's rhythms and intensity. In each instance, it seems to make such good sense. During ancestral times, for example, we were drawn to foods with high fat and sugar contents. We learned not to say no when such foods were present, as in honey from a beehive. But these foods were not present often. Today they are. Our fast foods are high in fats and sugars, and available 24/7, and it is difficult for many people now to say no to them, even though these foods harm people's physical and mental health. Many of us enjoy eating strawberries when they come into season during the summer months. Why stop there? We do not. We now import strawberries grown across the world. Unfortunately, they have lost a good deal of their original intense flavor. They have nice color, though. Some of us enjoy staying up and writing late into the night. That is a pleasure made possible by electric lights. The same lights blind us to the night sky. And so it goes for so many of our foods, conveniences, and artifacts—something gained, but something lost. We are usually aware of the gains and lose track of the losses.

Against this backdrop, here is a reflection that we presented to close one of our talks on rewilding at an environmental conference (Kahn & Hasbach, 2009):

There's a mountain pool that you find hiking up the wild river. The water emerges into it from porous volcanic rock. The water flows from the cold country. It's too cold to plunge in. But you're in. You're in because your lover is nearby and you need to prove your manly hood. But just as fast you're out. Your mind can't believe that mere water can be that cold. It would have been twenty strokes across. You give up that thought. You need to get warm fast. How? It's easy. You move naked to your beloved and put your arms around each other. Other people say that that pool is too cold. You can't swim in it. They say let's make a better one. They do. It's filled with chlorinated water that's not too hot and not too cold. Every day of the year it's that same temperature. It's called the tepid pool. The pool man comes once a week. He squeegees the sides and adds blue dye and oils the pump. The pool man says you gotta love the tepid pool.
  Where would you like to swim? Your choice.

Flattening the Hierarchy: Women, Bodies, and Brains

For more than six thousand years, our social relations with one another have been influenced by a hierarchical structure of domination. One of the central forms of such domination—although by no means the only— has been men dominating women. It is called patriarchy. Riane Eisler (1987, p. xix) says that patriarchal societies are not only "rigidly male dominant" but also "have a generally hierarchic and authoritarian social structure and a high degree of social violence, particularly warfare."

Most religions establish and sanction patriarchal worldviews. The Bible does so from its genesis. As even young children know, the Bible says that God created man. After that

the rib which the Lord God had taken from the man he made into a woman and brought her to the man. Then the man said, "This at last is bone of my bones and flesh of my flesh; she shall be called Woman, because she was taken out of Man." (Genesis 2:22–23, revised standard version)

Just like that, a male god through man strips woman of her most potent expression of feminine power: the ability to create and bring forth new life.

The Bible then teaches us that woman errs easily (Satan successfully tempts her). It teaches us that woman herself is a temptress (she seduces Adam) and through her failings brings woe on the world. The Bible teaches us that a woman's sexuality is sinful, that she has poor judgment and is intellectually inferior to man, and that it is her place to obey man (Stone, 1976). In I Timothy (2:11–14), for example, it is written:

Let a woman learn in silence with all submissiveness. I permit no woman to teach or to have authority over men; she is to keep silent. For Adam was formed first, then Eve; and Adam was not deceived, but the woman was deceived and became a transgressor.

This passage could be titled: *The Woman Did It.*

In I Corinthians (11:3, 7, 9) it is written:

For man is not of the woman, but the woman of the man. Let the woman keep silence in the churches, for it is not permitted unto them to speak; but they are commanded to be under obedience, so saith the law. And if they learn anything, let them ask their husbands at home; for it is a shame for women to speak in the church.

This passage could be titled: *Woman, Be Quiet.* We could similarly make our way through an entire reading of this patriarchal text.

There are many people who believe that the Bible or other male-dominant, hierarchical religious texts—such as the Koran or

Bhagavad-Gita—either speak truths from a divine source ("fundamental-ists" are of this camp) or represent truths of the natural order: the way it has always been for humans as we evolved into who we are today. But male-dominant, hierarchical religious systems are actually quite recent in the evolution of our species.

These systems began to emerge after agriculture replaced hunter-gath-erer life about ten thousand years ago. In the process, food production increased. That allowed families to settle into a place. Both made it pos-sible to have larger families. At that junction, two social structures were then open for the development of human culture. The first, the village, comprised "voluntary co-operation, mutual commendation, wider com-munication and understanding" (Mumford, 1961, p. 89). According to Eisler (1987, p. xvii), such a social system was "primarily based on the principle of linking rather than ranking. . . . In this model—beginning with the most fundamental difference in our species, between male and female—diversity is not equated with either inferiority or superiority." Eisler proposes that the original direction for mainstream culture was toward partnership, but that following a period of chaos and almost total cultural disruption, there occurred a fundamental social shift. She ties the dawn of patriarchy to the appearance of the nomadic bands of pastoral herders who worshipped a warrior god and ushered in a very different form of social organization some seven thousand years ago. These bands then solidified into what Jared Diamond (1996) calls chiefdoms, which involved a hierarchical social structure that monopolized force in a cen-tralized agency to establish its rule. In chiefdoms, for the first time in hu-man history, religions emerged that justified kleptocracy: the transfer of net wealth from commoners to the upper classes. And in chiefdoms, chiefs were men. It is still that way today.

It was never that way in our hunter-gatherer life. Rather, then, humans honored the life-giving, life-sustaining abilities of women and the strength skills of men within a relatively flat hierarchical social structure (Eisler, 1987; Gimbutas, 1982; James, 1957; Stone, 1976; Mellaart, 1975).

In reference to Neolithic imagery that indicated an understanding of the joint roles of women and men in procreation, Eisler (1987, pp. 26–27) writes that

while the feminine principle as the primary symbol of the miracle of life perme-ated Neolithic art and ideology, the male principle also played an important role. . . . All this imagery reflects the markedly different attitudes prevailing in the Neo-lithic about the relationship between women and men—attitudes in which linking rather than ranking appears to have been predominant.

Eisler (ibid., p. 27) quotes archaeologist Marija Gimbutas,

The world of myth was not polarized into female and male as it was among the Indo-Europeans and many other nomadic and pastoral peoples of the steppes. Both principles were manifest side by side. The male divinity in the shape of a young man or male animal appears to affirm and strengthen the forces of the creative and active female. Neither is subordinate to the other: by complementing one another, their power is doubled.

Thomas (2006) provides important characterizations of what a comparatively flat and nonpatriarchal social system looks like in hunter-gatherers from her experience in the 1950s living with the Ju/wasi Bushmen. According to Thomas, this group at that time was living largely as humans had lived since the Upper Paleolithic period, some thirty-five thousand years ago. In Thomas's account, though gender differences were evident in the activities performed by the Ju/wasi, the roles of men and women were viewed as equally essential and equally respected. Thomas (ibid., p. 175) writes:

In most ways, women were the equals of men, fully as respected, fully as important in decision making, fully as free to choose a spouse or get divorced or own a n'ove. . . . Men also were the equals of women, fully as tender toward their children, fully as ready to take part in daily tasks such as getting water or firewood. Yet there was a great dividing line between men and women that the Ju/wasi did not cross. . . . The division came down to childbearing and hunting. Matters of birth were only for women, and matters of hunting were only for men.

These key role distinctions seem to follow the primal energies of sexuality and aggression. On the role distinction of women, Thomas (ibid., p. 175) states, "It was the biological role of women that created this division. The act of life giving had so much power that men needed protection from it." She describes how a menstruating woman would separate herself from others in order to protect them from "the enormous power that clung to her at that time" (ibid.). Likewise, after giving birth, a new mother would bury all birth matter and mark it so that men could avoid it.

Perhaps referring to primal aggression, Thomas speaks of men's "natural bravery" related to the hunt and the many protection-related roles taken on by men. She describes how they confronted lions that wandered into camp, greeted human visitors, and accompanied women on overnight trips in order to protect, not supervise, them. She writes, "They hunted the world's most dangerous game with quarter-ounce arrows, they stood off lions and dealt with strangers, all without a shred of bravado or machismo that so characterizes the men of other societies, including ours" (ibid., p. 176).

Thomas discusses how the Ju/wasi women did not engage in sex with a hunter before the hunt, take part in the hunt, or even touch hunting equipment for fear of harming the hunter or the success of the hunt. She points out that today, such prohibitions are thought of as indicators of the inferiority of women. But she is clear that this was not true for the Ju/wasi. She notes, "The power of men was fragile and required direction and development. The power of women was strong and exuded from them naturally, from the day they reached the menarche" (ibid.). Thomas makes the point that neither menstruation nor ovulation requires skill, and if pregnancy occurred, childbirth will follow with or without experience. By contrast, hunting required much skill and knowledge, experience and hard work. She observes that "perhaps the passive power of women was the stronger of the two, but the active power of men was more apparent" (ibid.).

According to Thomas, whatever inequalities may have existed between Ju/wasi men and women, these did not apply to marriage or divorce practices. They conceptualized marriage as a rite of passage, and once in the married state, they usually remained there, if not always with the original partner. Although relatively rare, polygamy did occur, and was allowed for both men and women. Divorce was easily achieved by making an announcement, and it was equally available to and utilized by men and women alike.

Like other hunter-gatherer cultures, the Ju/wasi practiced infanticide. If children were born less than three or four years apart, both children were at risk for survival. "She could nurse one baby [at a time] to self-sufficiency, but she couldn't nurse two" (ibid., p. 196). Thomas maintains that children of both sexes were equally welcomed, so it was the spacing of children as opposed to their gender that determined their fate. The mother held the power to make the difficult decision of whether to put down a newborn child. If the mother decided it needed to be done, she was the person who did it.

Our point is relatively simple. We are saying that hierarchical systems that thrive through the domination of people over other people, and particularly of men over women, and that are reified in most, if not all, of the major religions of the world today, do not reflect a natural order. They do not represent an essential quality of human nature. They did not exist for tens and even hundreds of thousands of years in our evolutionary history. If anything, these systems represent a perversion of primal energy. These systems need to be changed. It is not a matter of going back to the social structure of hunter-gatherer times because we no longer are hunter-gatherers, and our populations involve not hundreds of people in a tribe but instead billions of people on a planet. But the solution, in no small way, involves flattening the social hierarchy, which includes re-envisioning, valuing, and affirming feminine principles and energy.

Toward such re-envisioning, the biblical "fall" can be flipped over and seen anew. In *Paradise Lost*, Milton (1674/1978) does just that, although he probably did not know it. In his epic account of Genesis, Milton (ibid., book 1, 25) seeks to "justify the ways of God to men." Mostly, he makes God look bad, he makes Satan intriguing, and he embodies Eve with many essential and laudatory feminine qualities. Eve, for example, is taken with Satan's argument that knowledge is good, and that a God that prevents humans from having knowledge reifies an unjust hierarchical structure. Satan asks Eve, "Why then was this [knowledge] forbid? Why but to awe, / Why but to keep ye low and ignorant / His worshippers" (ibid., book IX, 703–705). After Eve eats from the fruit of the tree of knowledge, she wonders what to do next. In part, she is intrigued with the idea that she is now more powerful than Adam. That has an initial appeal to her identity, and she wonders if perhaps Adam would find it alluring: "So to add what wants / In Female Sex, the more to draw his Love, / And render me more equal, and perhaps, / A thing not undesirable, sometime / Superior" (ibid., book IX, 821–825). But she checks herself: "For inferior who is free" (ibid., book IX, 825). She thus reasserts her desire for a partner in knowledge: "But keep the odds of Knowledge in my power / Without Copartner?" (ibid., book IX, 820–821). She goes on to say that "so dear I love him, that with him all deaths / I could endure, without him live no life" (ibid., book IX, 832–833).

So Eve decides to tell Adam. She has some explaining to do. She tells Adam that knowledge makes her more alive, more fully herself, and that she cannot imagine a life where she does not share knowledge of good and evil with the man she loves: "Th' effects to correspond, opener mine Eyes, / Dim erst, dilated Spirits, ampler Heart, / And growing up to God-head; which for thee / Chiefly I sought, without thee can despise" (ibid., book IX, 875–878). Adam is then "seduced." Yet in Milton's hand the seduction is not altogether unpleasing.

She embrac'd him, and for joy
Tenderly wept, much won that he his Love
Had so ennobl'd, as of choice to incur
Divine displeasure for her sake, or Death.
In recompense (for such compliance bad
Such recompense best merits) from the bough

She gave him of that fair enticing Fruit
With liberal hand: he scrupl'd not to eat
Against his better knowledge, not deceiv'd,
But fondly overcome with Female charm.
(ibid., book IX, 990–999)

The primal sexuality of woman brings a unique vitality and powerful energy to relationships—one with which man is often smitten. It motivates an appreciation of strength and beauty. It encompasses the subtleness of woman's desire and mystery of seduction. Perhaps it also fuels the unquenchable thirst for knowledge along with wholeness and equality.

During the early time of agriculture, when the feminine principle thrived, the pleasures of sex were experienced as sacred, and the woman's body was imaged as a magical vessel. As Eisler (1995, p. 25) writes:

They must have observed how it [the woman's body] bleeds in rhythm with the moon and how it miraculously produces people. They must also have marveled that it provides sustenance by making milk for the young. Add to this woman's seemingly magical power to cause man's sexual organ to rise and the extraordinary capacity of woman's body for sexual pleasure—both to experience it and to give it—and it is not surprising that our ancestors should have been awed by woman's sexual power . . . [nor is it surprising] that the sexual union between woman and man, the source of life, love, and pleasure, should have been for our Paleolithic and Neolithic ancestors an important mythical-religious motif.

In contrast, patriarchal religions image the woman's body as sinful. Christianity, in particular, uses the story of original sin to seal the deal on a hierarchical order of men over women.

Still, today, the male-dominated church, political systems, and medical communities seek to usurp a woman's power over such issues as birth control and abortion. Still, today, men in fundamentalist societies sanction and perform clitoridectomies to obliterate woman's sexual pleasure. Granted, many women in developed countries have come a long way in gaining access to knowledge through full participation in institutions of higher learning. Likewise, there are increasing numbers of women holding positions of power in fields previously reserved for men. But women are not so free to bring their unique vitality, beauty, and feminine power and energy into these traditionally male-dominated realms. For still, today, men the world over seek to harness female sexuality and denigrate its expression into narrow, stereotyped polar roles of Madonna or whore in order to gain control of feminine power. Too, some women, at least in part, have so bought into a patriarchal worldview that they have become their own enforcers.

Feminine power and sexuality can be dangerous. And alluring. Male power and aggression can be, too. They are both part of the primal energy that we spoke of earlier. We asked then, in the context of an excerpt from Baldwin's *Another Country*, what the sexual primal energy between man and woman might look like without violence and domination. We proposed that the answer might involve the concept of moral reciprocity based on responsible use. We offered there what was essentially a male version of that moral reciprocity. Milton's Eve captures a female version. Both the male and female versions are beautiful, dangerous, tempting, and healthy. They are not original sin.

Toward the end of Genesis (3:16), we learn of God's punishment to Eve:

To the woman he said, "I will greatly multiply your pain in childbearing;
In pain you shall bring forth children,
Yet your desire shall be for your husband, and he shall rule over you.

Eisler (1995, p. 25) points out, "Most dramatically, it [this punishment] requires an almost total reversal and vilification of precisely what was once revered: nature, sex, pleasure, and above all, the life-creating-and-sustaining sexual power of woman." Although Milton tried to vilify Eve, because that was his theological canon, he ended up in many respects portraying a primal woman with body and brains—one who is smart, powerful, beautiful, and seductive, and interested in a partnership with others.

Technology

We said at the outset that we aim for a proactive agenda—one that is as practical as it is future oriented—and that we would integrate our views on the rewilding of the human species with what we take to be a truism: that we are not only a natural but also a technological species. From digging sticks and stone axes to bulldozers and digital computation, technology has been adaptive and forms a part of our selves. We cannot survive without technology. As the decades unfold, we will not be able to survive without increasingly pervasive, complex, and computational forms of technology. It is also worth noting that because it provides largely egalitarian access to information and communication, technology (such as the Internet) can help to flatten social hierarchies. What, then, is the role of technology in our proposal for the rewilding of the human species?

Although our technologies often benefit us as a species, they have costs that are not so readily apparent, especially when it comes to our encounter with the wild. Let us consider a relatively simple technology: a flashlight. Imagine being in the dark woods. A flashlight brings a measure of security insofar as we can easily find things, if we happen to be camping, or we can easily find our way, if we happen to be walking. Perhaps this measure of security allows us to overcome our fear of being in the woods at night, and thus even makes possible the night encounter.

Yet the flashlight also limits our experience of the wild. It prevents our eyes from adapting to the dark. When walking, it prevents us from picking up shadowed forms of landscape by which to find our way. It limits us from using our feet to feel the texture of the land to aid in our navigation—where, perhaps, we recognize a rock-strewn stretch or particular spot of meadow grass. It limits us from using our bodies to feel the contours of the land—where, perhaps, we recognize a slight rise on an otherwise-flat path. It limits our ears from picking up the nightlife— perhaps a deer or owl—because such life is more likely to have already fled. The flashlight prevents us from seeing and more fully experiencing the night sky in its most dazzling arrays, and in and through its vastness, experience awe, humility, and grandeur.

Virtually all technologies are of this form. GPS navigation devices provide many people access into the wild. But to navigate the land these devices demand eyeballs on digital screens, and thus come at the cost of attending to the land's wild features. Cars and roads offer comparatively quick access into wild areas. Such access, though, comes at the cost of longer and wilder outings, including the longer rhythm of movement away from human settlement and the return. Technologies allow us to build houses to shield us from the wild forces of cold, rain, wind, and snow. Technologies allow us to build many houses, and the infrastructure to place them side by side and even one on top of another, creating large infrastructures, called cities, which eradicate wild land and wild animals from our daily lives. In our cities, the single flashlight is multiplied perhaps a billionfold or more. The result is light pollution. It prevents people from ever seeing the night sky even when they turn off their own individual lights. As technologies become ever more efficacious, and as nature stays the way it has largely been, it takes less and less time, and fewer and fewer people, to make inroads into the wild and destroy it.

We asked what is the role of technology in our proposal for the rewilding of the human species. The first part of our answer is that even as we can use technology to gain access to the wild, which is a plus, the same technology often leads us to control, dominate, and destroy the wild, or otherwise engage it less deeply. The second part of our answer recognizes that we can use technology to display and simulate the wild. That happens readily through long-established mediums of magazines (e.g., *National Geographic*), television, and movies (e.g., on the Discovery Channel). More recently, video games and digital worlds, such as Second Life, move us increasingly to virtual realities that simulate wild places and wildlife. *Star Trek*'s *holodeck* represents a future vision of such a virtual reality.

Can such representations and simulations, now and in the future, substitute in some meaningful way for experiences with actual wildness? An emerging body of research by Kahn and his colleagues bears on this question (Kahn, 2011; Kahn, Friedman, Gill et al., 2008; Kahn, Friedman, Perez-Granados, & Freier, 2006; Melson et al., 2009). Over the last decade, they have investigated the psychological effects of interacting with what they call *technological nature*: technologies that mediate, augment, or simulate the experience of the natural world. In one study, for example, HDTV-quality real-time views of a local nature scene were displayed on a fifty-inch plasma-display "window" in windowless offices of faculty and staff in a university setting (Friedman, Freier, Kahn, Lin, & Sodeman, 2008). The view through the plasma window included water in the foreground, as part of a public fountain area, and then extended to include stands of deciduous trees on one side, and a grassy expanse that allowed a visual "exit" on the other. This office view was chosen to include features that people usually find aesthetically pleasing and restorative in nature (Kaplan & Kaplan, 1989; Orians & Heerwagen, 1992). Data were collected over a sixteen-week period, which resulted in 652 pages of interview transcripts, journal entries, and responses to email inquiries. The results showed that participants enjoyed the plasma-display window, and benefited from it in terms of their psychological well-being, cognitive functioning, and connection to the natural world. Four weeks after the display window was removed from participants' offices, all of them felt less connected to the local area and would have liked to have had the plasma window reinstalled.

In another study (Kahn, Friedman, Gill et al., 2008), ninety participants in an office setting were exposed to one of three conditions. One condition was of the same plasma window and real-time technological nature view as in the previous study. In the second condition, the plasma window was removed; it had covered up an actual glass window that looked out to the actual nature scene that had been displayed by the plasma window. In other words, the glass window condition (condition 2) displayed the actual view that had been technologically mediated by the plasma window (condition 1). In the third condition, drapes completely covered the actual window, and thus the office was made in effect into an inside office without any view to the outside (actual or technological). Three key results emerged. First, in terms of heart rate recovery from low-level stress, working in the office environment with a glass window that looked out on a nature scene was more restorative than working in the same office without the outside view (the blank wall condition). Second, in terms of this same physiological measure, the plasma window was no different from the blank wall. Third, when participants looked longer out the glass window, they had greater physiological recovery; but that was not the case with the plasma window, where increased looking time yielded no greater physiological recovery.

Taking both studies together, the central finding was that the plasma nature window was better than no nature, but not as good as actual nature.

This central finding was corroborated in five other studies, which cut across two other forms of technological nature: an advanced robot dog (AIBO) and a Telegarden—an actual garden in Austria that allowed remote "gardeners" to plant and tend seeds by controlling a robotic arm through a Web-based interface. In one study, for example, children's and adolescents' interactions with and reasoning about AIBO were compared to their interactions with and reasoning about a biologically live dog (Melson et al., 2009). The results showed that children and adolescents more often affirmed that a live dog, in comparison to AIBO, had mental states (83.6 percent live dog, and 56.2 percent AIBO), could be a companion (91.4 percent live dog, and 70.0 percent AIBO), and had moral standing (86.3 percent live dog, and 75.7 percent AIBO). These numbers are intriguing in two ways. On the one hand, the numbers (and resulting inferential statistics) show that AIBO was not as compelling as a live dog. On the other hand, it is surprising that AIBO—a robot dog—was as compelling as it was.

It is important to note that the nature used in these technological nature research studies was of a domestic or pastoral form. The plasma-display window studies involved the real-time technological representation of water in a large public fountain, and expanses of grass and trees. The robot dog studies involved the technological embodiment of a biological dog. And the Telegarden study involved mediated access to an actual garden. What would have been found had the empirical studies on technological nature involved wild nature? Our guess is that the same dual-edged pattern would have emerged. On the one side—and a plus for technology—it is likely that some of the experience of wildness can be created, induced, and fostered through our technological creations. On the other side, it is likely that technological wildness cannot create, induce, or foster the human experience of wildness as effectively or fully as actual wildness.

In terms of the human relationship with nature, Kahn (1999, 2002, 2011) has suggested that one of the central psychological problems of our lifetime is *environmental generational amnesia*. The basic idea here is that each generation constructs a conception of what is environmentally normal based on the natural world encountered in its childhood. The problem arises because with each generation, the amount of environmental degradation increases, but each generation tends to take that degraded condition as the nondegraded condition, as the normal experience. Therefore the baseline shifts downward for what counts as rich, healthy, and diverse nature. For instance, in a study of African American children's environmental views and values in Houston, Texas (Kahn & Friedman, 1995), it was found that a significant number of the children understood the idea of air pollution, but they did not believe that Houston had such a problem, even though Houston was then (and still remains) one of the most polluted cities in the United States. In interpreting these results, it was suggested that these children lacked a comparative experiential baseline from areas with less pollution by which to recognize that Houston was itself polluted.

If technological wildness continues to become more sophisticated and pervasive, and if we as humans continue to degrade and destroy wild places—and both situations seem almost certain to occur—then there will be a downward shift, as there has been already, in the baseline across generations for what counts as wild.

This form of environmental generational amnesia will continue to make the rewilding of our species difficult. For people will not easily believe that in losing wildness, they have lost anything of importance. As a comparison, imagine people who have grown up without exposure to music. Imagine that we tell them that they are missing a means of expression along with a way of experiencing the world that is uniquely powerful and beautiful in human life. It is likely that they would respond that they are not missing anything by not having this thing called music. For they would likely not miss what they have never known. We hope that it does not get to that point with wildness in human lives.

Conclusion

The hunter-gatherers had it hard and good. But we cannot go back to that life. The world is too populated and environmentally degraded. And as a species, through adaptation, we too have changed, and are now too dependent on agriculture, too populated, too urbanized, and too woven into our advanced technological infrastructures. The bad news is that we are not doing so well as a species. The good news is that, if we choose differently, we can have it better than our ancestors did. Why? Because though the world has somewhat changed, and we have too, the need for wildness without and within remains an essential quality of human nature. We can no more change that than we can change the fact that to thrive we need clean air to breathe, physical exercise, and trusting and loving relationships in our lives. We can have it better than our ancestors did if we rediscover our ancestral wildness, and integrate many of those qualities and experiences into the structure of our scientific and technological selves.

Toward this end, we have offered a beginning account of the rewilding of the human species. Our account depends, in part, on interaction with a wild nature "out there" in the external world. Land, for example, that is big, vast, untrammeled, and not managed. Natural systems that are self-organizing, self-regulating, and generative based on their own authority. And animals that are neither captive nor domestic. To access this wild nature and go deeper within it, we often need to act in ways that seem contradictory but that are not. We need, for instance, at times to relinquish control. That does not lead to our being out of control but rather to allowing ourselves, like the boy in Faulkner's *The Bear*, to fit more readily into a larger self-regulating system of the more-than-human world. We need to be open to our fear of the wild. That does not mean being paralyzed by it; just the opposite. It means being more open, quietly alert, and present. Or in those times of immediate threat, which come less frequently in nature than our digital portrayals of nature convey, it means being confident, strong, and brave.

Rewilding also involves valuing and finding healthy expressions of the wild within. We emphasized primal passions: sexual and even aggressive. Aggression is the more problematic of the two. It usually takes forms that are destructive and in our view perverted. When our species moved outward from Africa some thirty-five thousand years ago and then, in land from which we did not coevolve, moved into an agriculture life, populations increased. Hierarchical social systems emerged. And unfortunately, the warrior ethos took hold and exists today as the dominant worldview.

Our challenge is to reject it not by repression but instead by tapping more directly its primal—"original"—energy and functions.

As such, we argued that we need a flatter hierarchy in social systems so as to minimize control and domination. We emphasized that women have been subject to such control and domination, especially of their sexuality and quest for knowledge, ever since the warrior ethos took hold and was reified within the major religions of the world today. We sought to reinstate the place of a woman with body and brains: one who is smart, powerful, beautiful, and seductive, and interested in a partnership with rather than being dominated by or having domination over others.

Finally, we spoke of how we are not only a natural but a technological species as well. The technologies we create and bring into our lives are frequently a tremendous boon. Technologies can also provide access into wild nature, and represent, simulate, and mediate it. These uses can be good. But we have also highlighted that much more often than not technologies allow us to control nature, which is at odds with experiencing the wild.

That is one of the lessons that the boy in Faulkner's story learned. To experience the wild, he had to relinquish technological control. He had to cast aside his compass and watch. After that, he became lost farther out in the wild than he had ever traveled. Recall that only then did he see the crooked print in the wet ground filling with water, and another above it, and another, and another. . . . Then, and only then, did the boy see the bear—the bear that was

too big for the dogs which tried to bay it, for the horses which tried to ride it down, for the men and the bullets they fired into it; too big for the very country which was its constricting scope. (Faulker, 1961, p. 191)

The boy saw the bear, and the bear saw the boy, and the latter happened far earlier than the former—and after the bear looked back at the boy across one shoulder, and after it was gone, Faulkner ends this section of his story with a surprising metaphor. He writes that the bear "faded, sank back into the wilderness without motion as he [the boy] had watched a fish, a huge old bass, sink back into the dark depths of its pool and vanish without even any movement of its fins" (ibid., pp. 202–203). How could a fish compare to a bear, let alone the biggest and most wild bear of them all? Not easily. But the passage works because it conveys another primal aspect of the wild, the movement back into mystery, into the "dark depths" from which the Other comes, and the mysterious primal parts of ourselves, which we can never fully know.

References

Applebome, P. (2009, September 7). An idyllic pool becomes the scene of a battle. *New York Times*, p. A12.

Baldwin, J. (1960). *Another country*. New York, NY: Vintage.

Diamond, J. M. (1996). *Guns, germs, and steel: The fates of human societies*. New York, NY: W. W. Norton.

Eisler, R. (1987). *The chalice and the blade: Our history, our future*. New York, NY: Harper and Row.

Eisler, R. (1995). *Sacred pleasure: Sex, myth, and the politics of the body*. New York, NY: HarperCollins.

Faulkner, W. (1961). *The bear*. New York, NY: Vintage.

Friedman, B., Freier, N. G., Kahn, P. H., Jr., Lin, P., & Sodeman, R. (2008). Office window of the future?—Field-based analyses of a new use of a large display. *International Journal of Human-Computer Studies*, 66, 452–465.

Gimbutas, M. (1982). *The goddesses and gods of Old Europe, 7000–3500 B.C.* Berkeley, CA: University of California Press.

Gullone, E. (2000). The biophilia hypothesis and life in the 21st century: Increasing mental health or increasing pathology? *Journal of Happiness Studies*, 1, 293–322.

James, E. O. (1957). *Prehistoric religion*. New York, NY: Barnes and Noble.

Kahn, P. H., Jr. (1999). *The human relationship with nature: Development and culture*. Cambridge, MA: MIT Press.

Kahn, P. H., Jr. (2002). Children's affiliations with nature: Structure, development, and the problem of environmental generational amnesia. In P. H. Kahn, Jr., & S. R. Kellert (Eds.), *Children and nature: Psychological, sociocultural, and evolutionary investigations* (pp. 93–116). Cambridge, MA: MIT Press.

Kahn, P. H., Jr. (2011). *Technological nature: Adaptation and the future of human life*. Cambridge, MA: MIT Press.

Kahn, P. H., Jr., & Friedman, B. (1995). Environmental views and values of children in an inner-city black community. *Child Development*, 66, 1403–1417.

Kahn, P. H., Jr., Friedman, B., Gill, B., Hagman, J., Severson, R. L., Freier, N. G., Feldman, E. N., Carrère, S., & Stolyar, A. (2008). A plasma display window?— The shifting baseline problem in a technologically-mediated natural world. *Journal of Environmental Psychology*, 28, 192–199.

Kahn, P. H., Jr., Friedman, B., Perez-Granados, D. R., & Freier, N. G. (2006). Robotic pets in the lives of preschool children. *Interaction Studies: Social Behaviour and Communication in Biological and Artificial Systems*, 7, 405–436.

Kahn, P. H., Jr., & Hasbach, P. H. (2009, November 14). *The rewilding of the child: Toward a research agenda*. Paper presented at the annual meeting of the North American Association for Environmental Education, Portland, OR.

Kahn, P. H., Jr., Saunders, C. D., Severson, R. L., Myers, O. E., Jr., & Gill, B. T. (2008). Moral and fearful affiliations with the animal world: Children's conceptions of bats. *Anthrozoos*, 21, 375–386.

Kaplan, R., & Kaplan, S. (1989). *The experience of nature: A psychological perspective.* New York, NY: Cambridge University Press.

Marshall, L. (1976). *The !Kung of Nyae Nyae.* Cambridge, MA: Harvard University Press.

Mellaart, J. (1975). *The Neolithic of the Near East.* New York, NY: Scribner.

Melson, G. F., Kahn, P. H., Jr., Beck, A. M., Friedman, B., Roberts, T., Garrett, E., Gill, B. T. (2009). Children's behavior toward and understanding of robotic and living dogs. *Journal of Applied Developmental Psychology, 30,* 92–102.

Milton, J. (1978). Paradise lost. In M. Y. Hughes (Ed.), *Complete poems and major prose* (pp. 173–470). Indianapolis, IN: Odyssey Press.(Original work published 1674).

Muir, J. (1976). *The wilderness world of John Muir.* E. W. Teale, (Ed.). Boston, MA: Houghton Mifflin.

Mumford, L. (1961). *The city in history.* New York, NY: Harcourt.

Nelson, R. (1989). *The island within.* New York, NY: Random House.

Orians, G. H., & Heerwagen, J. H. (1992). Evolved responses to landscapes. In J. H. Barkow, L. Cosmides, & J. Tooby (Eds.), *The adapted mind: Evolutionary psychology and the generation of culture* (555–580). New York, NY: Oxford University Press.

Primal. (2011). *Merriam-Webster's online dictionary.* Retrieved from http://www.merriam-webster.com/dictionary/primal

Rolston, H., III. (1989). *Philosophy gone wild.* Buffalo, NY: Prometheus Books.

Shepard, P. (1993). On animal friends. In S. R. Kellert & E. O. Wilson (Eds.), *The biophilia hypothesis* (pp. 275–300). Washington, DC: Island Press.

Shepard, P. (1998). *Coming home to the Pleistocene.* Washington, DC: Island Press.

Stone, M. (1976). *When god was a woman.* New York, NY: Harcourt Brace.

Thomas, E. M. (2006). *The old way: A story of the first people.* New York, NY: Farrar, Straus and Giroux.

Ulrich, R. S. (1993). Biophilia, biophobia, and natural landscapes. In S. R. Kellert & E. O. Wilson (Eds.), *The Biophilia hypothesis* (pp. 73–137). Washington, DC: Island Press.

Name Index

Valentine, P., 169
Varela, Francisco J., 37
Varga, D., 96–97, 103
Vest, Jay Hansford, 185
Vickery, J.A.G., 190
Votier, S. C., 85
Vygotsky, Lev Semyonovich, 101

Waldron, S., 85
Waley, Arthur, 165
Walker, J. S., 16
Walker, R. N., 108
Wallace, L. L., 16
Walls, Laura Dassow, 46
Wamba, K., 128
Wamba, N., 128
Wamba, P., 128
Wania, F., 88
Watt, Rob, 10–11
Waxman, S., 106
Weaver, J. L., 19
Wei-ming, Tu, 41
West, Paige, 172
White, E. B., 109
White, P. N., 109
Whitman, Walt, 141
Whittingham, L. A., 75
Widtsoe, J., 189
Williams, Michael, 81
Williams, Terry Tempest, 1, 3, 6, 46
Wilson, E. O., 6, 33, 37, 46, 93, 104,
    148, 185
Wingfield, J. C., 83
Winnebago, 168–169
Wittgenstein, Ludwig, 32, 36, 43
Wright, H. F., 108

Yeldham, C., 170
Young, S. P., 5

Zalik, N. J., 86
Zhuangzi, 35, 164
Zimmer, C., 39, 44
Zolbrod, P., 177

Subject Index